The Search for Extraterrestrial Intelligence

LISTENING FOR LIFE IN THE COSMOS

THOMAS R. McDONOUGH

*To Vince —
With best wishes, and
hopes that we hear from
E.T. soon!
Tom McDonough, 8/10/91*

WILEY SCIENCE EDITIONS

JOHN WILEY & SONS, INC. *New York • Chichester • Brisbane • Toronto • Singapore*

Publisher: Stephen Kippur
Editor: David Sobel
Managing Editor: Katherine S. Bolster
Production Service: G&H SOHO, Ltd.
Designer: Laura Ferguson

Library of Congress Cataloging-in-Publication Data

McDonough, Thomas R.
  The search for extraterrestrial intelligence.

  Bibliography: p.
  1. Life on other planets. I. Title.
QB54.M527 1987   574.999   86-15905
ISBN 0-471-84684-8

Printed in the United States of America

87 88 10 9 8 7 6 5 4 3 2

PERMISSIONS

I WISH TO THANK Davis Publications, Inc. for permission to reprint with modifications my article *They're Trying to Tell Us Something* from the March and April, 1969, issues of *Analog* magazine, in Chapter 9. (© 1969 by The Condé Nast Publications, Inc.)

Grateful acknowledgment is also made to the following for permission to use the illustrations appearing on the pages listed below.

Pages 7, 8, 9, 58, 61, 74, 77, 79, 81, 82, 83, 84, 85, 86, 87, 88, 90, 91, 92, 93, 94, 95 (left), 97, 98, 99, 100 102, 103, 107 (bottom), 108, 109, 115, 123, 146, 147 (top), 157, 158, 210, 211, 212, 213, and all color photos except those listed below: Courtesy JPL/NASA.

Pages 15, 56 (bottom), 59, 60 (left), 125, 127, 153: © 1986 Sidney Harris.

Page 19: "The Far Side," by Gary Larson. © 1985 Universal Press Syndicate. Reprinted with permission. All rights reserved.

Page 22: "Fellows of the Zoological Society" by George Cruikshank, from *The Comic Almanack* (1851).

Page 23: Courtesy D. K. Yeomans.

Pages 25 (both), 26, 27: Lowell Observatory photographs. Courtesy Lowell Observatory.

Page 32 (left): © *Galaxy* magazine.

Page 32 (right): © 1976 David A. Hardy and Mercury Press, Inc. Reprinted from *The Magazine of Fantasy and Science Fiction*.

Pages 36, 37, 38, 39, 40, 41, 44: © Joel Hagen 1982, 1983, 1984, 1985.

Page 48: © 1966 Paramount Pictures Corporation. All rights reserved. Courtesy of Paramount Pictures.

Page 50: © by Universal Pictures, a Division of Universal City Studios, Inc. Courtesy of MCA Publishing Rights, a Division of MCA Inc. E.T.™ is a trademark of and licensed by Universal City Studios, Inc.

ACKNOWLEDGMENTS

MANY THANKS are due to my editor, David Sobel, and Katherine Bolster, managing editor, and also to Andrew Hoffer, Dawn M. Reitz, and to Julie Glass, my developmental editor, for enabling the evolution of the manuscript from the primeval ooze in the author's brain to the sleek creation you see before you. And special thanks are due my agent, Sharon Jarvis, for serving as the catalyst in this process.

I wish to especially thank Ray Bradbury for his time and for permission to quote from *The Martian Chronicles.*

I wish to thank the following for their help: Toni Acker (Wonder Works Studio), James Aulenback, J. Kelly Beatty (*Sky and Telescope*), Dr. Valentin Boriakoff (Cornell), Ben Bova, Bruce Bruhn (USPS), Greg Chang (Crown Books), Dennis M. Cole (Lockheed), Diane Cole, Charles R. Colgan (Scripps Institution of Oceanography), Sue Couchman (Crown Books), Dr. Michael M. Davis (Arecibo Radio Observatory, National Astronomy and Ionosphere Center, Cornell University), Prof. Raymond Davis, Jr. (University of Pennsylvania), Robert Dixon (Ohio State University), Dr. Frank Drake (U.C. Santa Cruz), Pamela Fields, Dr. Robert L. Forward (Hughes Aircraft Company Research Laboratories), Dr. Louis Friedman (The Planetary Society), Mike Gentry (NASA/JSC), Mel Gilden, Dr. Donald Goldsmith, Cheryl S. Gundy (Space Telescope Science Institute), Geoffrey Haines-Stiles, Dr. Norman Horowitz (Caltech), Dr. Paul Horowitz (Harvard), Helen S. Horstman (Lowell Observatory), the late Dr. J. Allen Hynek, Philip J. Klass (*Aviation Week*), Jon Lomberg, Julie Moskowitz (Amblin Entertainment), Dr. Lloyd Motz (Columbia), Dr. Bruce Murray (Caltech and The Planetary Society), Mary Jeanne Pickett (Photo Researchers, Inc.), Jonathan V. Post, Senator William Proxmire, Jeane Rae, Dr. Nancy Rallis (Boston College), Gene Roddenberry, Raymond T. Rye II (Smithsonian Institution), Dr. Carl Sagan (Cornell and The Planetary Society), Steven Spielberg

(Amblin Entertainment), Bug Sutton (Photo Researchers, Inc.), Neil Strum (Paramount Pictures Corp.), Sandy Taylor, and Margaret B. Weems (NRAO).

Thanks to the following JPL people for much help: Dr. Edward Belbruno, Frank Bristow, Joel K. Harris, Dr. Michael J. Klein, Dr. Thomas B. H. Kuiper, Joseph W. Stockemer, Sherry Wheelock, James Wilson, Jurrie van der Woude, and Dr. Donald K. Yeomans.

Thanks to Dr. Robert L. Forward and the Hughes Aircraft Company Research Laboratories for permission to reprint their figures; and to Dr. Louis Friedman, Charlene H. Anderson, Lyndine McAfee, and The Planetary Society for travel support, tape recordings of the META Symposium, and much other assistance. Many of the quotations of Robert Forward, Paul Horowitz, Philip Morrison and Carl Sagan were taken from the META Symposium sponsored by the Society.

Thanks to Margaret and Anthony Ywoskus for their hospitality in Boston, and to Tony for very useful discussions about history. And special thanks to the Sophie T. McDonough Clipping Service.

CONTENTS

INTRODUCTION

CALLING ALL E.T.'S

FOR THOUSANDS OF YEARS, humans have believed there were beings in the sky. In the nineteenth century, Mars was widely thought to be civilized. In this century, we have explored most of the planets and have sent robot emissaries to Mars to search for life. Tantalizing clues have been found suggesting that life may exist elsewhere in the universe.

For two decades, a small, dedicated band of men and women has been searching for intelligent extraterrestrial radio signals. They have scraped up the funds for equipment, often fighting the ridicule of their colleagues and the short-sightedness of politicians.

They have persisted. They know that signals from another civilization could be passing through your body at this very moment just as ordinary radio signals do. Without the hard work of these pioneers, they would never be found.

If these scientists are successful, their discovery will be the most important in the history of the human species, bar none. The universe has been around for billions of years longer than Earth, so other civilizations may be vastly older than ours. They have most likely faced the same problems of war, disease, starvation and pollution that plague humanity today. Any civilization that survived very long most likely solved these problems. The answers may just be waiting for the right person to tune to the right signal at the right moment.

Scientists call this work the Search for Extraterrestrial Intelligence, SETI for short (rhymes with "jetty"). It sounds like science fiction, but SETI is real science. It is happening today. There's a very good chance that, in the next few years, the reader will see the first proof of extraterrestrial beings in the headlines of all the world's newspapers.

At the same time, SETI is in a race against the clock. The very technology that makes it possible to detect other civilizations is now clogging the skies with radio pollution; spy satellites, communication satellites and others are making it harder and harder to find signals from other worlds.

However, as of now, if E.T. is phoning, *we* are listening. This book tells who's listening, and why, where and how. It tells the comedy and the drama of the story, including the false alarms when we thought we had connected with extraterrestrials. And it suggests what we may find when we answer the greatest long-distance call in history.

E.T. Phone Earth!

THE NATURE OF THE SEARCH

It seems impossible, in a large field only one shaft of wheat to grow, and in an infinite Universe, to have only one living world.
Metrodorus of Chios

IN 1938, A RADIO STATION was playing the toe-tappin' rhythms of Ramon Raquello and his orchestra, when an announcer broke in, saying, "Ladies and gentlemen, we interrupt our program of dance music to bring you a special bulletin from the Intercontinental Radio News. At twenty minutes before eight, central time, Professor Farrell of the Mount Jennings Observatory, Chicago, Illinois, reports observing several explosions of incandescent gas, occurring at regular intervals on the planet Mars."

The station returned to the music, but was soon interrupted again, this time with an interview of astronomer Richard Pierson of the Princeton Observatory. The announcer said, "You're quite convinced as a scientist that living intelligence as we know it does not exist on Mars?"

The astronomer replied, "I should say the chances against it are a thousand to one."

After a few more questions, the announcer said, "Ladies and gentlemen, I shall read you a wire addressed to Professor Pierson from Dr. Gray of the National History Museum, New York. '9:15 PM eastern standard time. Seismograph registered shock of almost earthquake intensity occurring within a radius of twenty miles of Princeton. Please investigate. Signed, Lloyd Gray, Chief of Astronomical Division.'

"Professor Pierson, could this occurrence possibly have something to do with the disturbances observed on the planet Mars?"

3

"Hardly. . . . This is probably a meteorite of unusual size. . . ."

Later, we heard a report from the impact site, where hundreds of people surrounded a large metal cylinder half-buried in a pit. As they watched, the top of the cylinder slowly unscrewed and something tentacled crawled out, as large as a bear, glistening like wet leather.

"Wait!" shouted the announcer. "Something's happening!"

There was a hiss, then a hum that grew louder and louder.

"A humped shape is rising out of the pit," said the announcer excitedly. "I can make out a small beam of light against a mirror. What's that? There's a jet of flame springing from that mirror, and it leaps right at the advancing men. It strikes them head on! Good Lord, they're turning into flame!"

And thus the nation heard Orson Welles' Halloween broadcast of H. G. Wells' *War of the Worlds*. Six million people listened to it, and so realistically was it done that, according to one estimate, approximately a million people were seriously frightened. Many panicked.

Thousands cried, prayed, sealed their homes against poison gas, or fled to shelter. "I thought the whole human race was going to be wiped out," said one college senior who had raced in his car away from the reported invasion, pushing the accelerator to the floor.

Why were a million people so quick to believe the Earth had been invaded by Martians?

To answer that, we need a glimpse of how history brought us to that point, and where it is taking us now. This is the origin of SETI— the scientific search for extraterrestrial intelligence—and this chapter is a preview of coming attractions. It is a survey of where we've been, where we are now, and where we are likely to be in the future—a thumbnail sketch of SETI, and a sampler of the areas explored in greater depth in later chapters.

The History of Aliens

Even before the invention of writing, humans undoubtedly imagined aliens in the sky. They called them gods, and gave them names like Mars, Venus and Jupiter. They drew connect-the-dot pictures in the sky, showing images such as those of the mighty hunter Orion and the beautiful Andromeda.

Philosophers speculated more seriously about the possibility of life beyond Earth. Ancient Greeks wondered whether there might be other worlds, and if so, whether they might be inhabited by thinking beings.

It was not until after Columbus discovered the New World, however, that Copernicus first applied real science to the problem. He showed that the Earth is just one of several planets circling the Sun. Copernicus' ideas were nearly suppressed, for they challenged the early Christian view of the Earth as the center of God's creation. But then Galileo turned his telescope on the heavens and found that not only was Copernicus right, but there were four moons orbiting around Jupiter, making it a miniature of the solar system. And he saw that the planets of the ancients were not just wandering stars, they were spheres like the Earth.

Galileo and his contemporaries had to fight imprisonment,

A flier for a nineteenth-century SETI lecture. We are advised that the speaker "brings to the work mature years, correct tastes, profound learning, skilled and brilliant powers of oratory, with great personal magnetism . . ."

torture and execution before their ideas were accepted. But gradually, people came to realize the truth: the Earth was nothing but one planet among many.

So why shouldn't the other planets be inhabited?

Astronomers, philosophers and writers imagined beings living on the Moon, Jupiter, Mars and the other known worlds. They envisioned civilizations as alien to themselves as the Chinese and the Aztecs had seemed to European explorers.

Speculation about other planets became popular among philosophers, writers and theologians.

Take Me to Your Leader

Writers loved the idea of other worlds with exotic beings. At first, this provided a way of satirizing human culture by depicting societies on other worlds that differed only slightly from our own. But in the nineteenth century, fantasy mated with science and produced science fiction. Mary Wollstonecraft Shelley's *Frankenstein* took the new scientific ideas about life that would soon lead to Darwin's theory of evolution, and envisioned the creation of a living being by science as a nightmare.

Jules Verne applied his vivid imagination to the rapidly growing technology and envisioned trips to the Moon and other scientific wonders.

Astronomers discovered markings on Mars that seemed to be artificial canals, and deduced the existence of a vast, advanced civilization. "Martian" became almost synonymous with "alien extraterrestrial" and writers such as H. G. Wells could not resist depicting the inhabitants of that world.

So when, on the brink of World War II, Orson Welles reenacted *War of the Worlds* as a radio docudrama, the public was quite ready to believe in Martians.

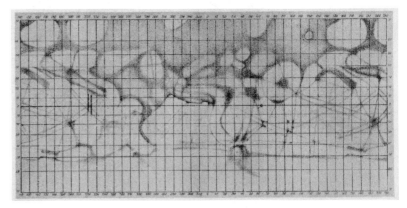

A Mars map by the nineteenth-century Italian astronomer Schiaparelli, one of those which convinced Percival Lowell that a canal-building civilization existed there.

Halley's Comet, as seen from Earth during its 1985–86 appearance. Comets appear to be dirty snowballs left over from when the planets formed.

Galactic Hitchhiker's Guide to Planets

Serious exploration of the idea of life on other worlds required understanding how our own life arose here. Darwin gave us the great theory of evolution which served as the framework for this problem, but many details were still unclear at that time.

The twentieth century solved many of these mysteries. We learned about the existence of the DNA molecule containing the blueprints of life in every creature and plant on Earth. We developed theories of how it all got started—about how that first reproducing molecule arose from inanimate chemicals. We learned enough to begin to speculate more intelligently about the possibility that similar processes might occur on other worlds.

Astronomers found many clues hinting that planets are common around other stars. Recently, for example, around some stars, rings have been seen that may be planets forming. Planets seem to be a normal byproduct in the formation of many stars in the galaxy.

The Real Invaders of Mars

Our own solar system provides us with nine known planets to compare, plus dozens of moons. Space travel, once the province of science fiction, has allowed us to probe many of them.

Mars, beloved of H. G. Wells and Orson Welles and countless other craftsmen of the imagination, turned out to be uninhabited. The canals of Mars are just channels, after all. But Mars is still a

VIKING LANDER 1   CAMERA 1 0.04   CE LABEL 11007/014
AZIMUTH 132.3/265.5   ELEVATION -20.22/ 0.22   CHANNEL MODE 13-2
DATA RATE 4000   TEMP 13C(23)   AIB RATE RCV
LINE 5-3024/26.6   BEGIN 0   RESCAN TOTAL 0
END 7 ELV 1.14.13   82.61 13
PIXEL FILE PTR 5   82.61   AT GMT 016/12.24.07
VCA038/ 2

Morning on the surface of Mars, as seen by the first *Viking* lander. Rocks, sand dunes and hints of distant craters are visible. The geology is similar to that of Death Valley. The large boulder at left is about ten feet wide. The diagonal foreground features near the center are part of the lander.

remarkable place, a world of canyons and volcanoes and craters bigger than any on Earth. And it does have water—a necessity for life as we know it—in the form of ice.

Mars could have once harbored life in ancient, balmier days, and there is a possibility that some form of life might yet exist in some cozy ecological niche underground or on the edge of the great polar ice caps.

Some of the other planets are similar to what we think Earth was like in the early days, billions of years ago, when life formed here. So they too could conceivably contain living creatures. And some of their moons have turned out to be surprising. An ice-covered moon of Jupiter is filled with a warm ocean; a moon of Saturn has an atmosphere thicker than ours, and may have seas of primitive organic chemicals.

Although extraterrestrial life has not been found in our solar system, the neighborhood has proven to be so full of unexpected sights that we may still have hope.

Is E.T. Phoning?

In 1959, the search for *intelligent* life on other worlds started a new era. Two scientists proposed looking for radio signals from other civilizations. More ideas once confined to science fiction became respectable science.

The next year, the first modern scientific search for radio signals from an extraterrestrial civilization began, and seemed to succeed almost immediately. Unfortunately, the signals turned out to be interference, but the most powerful tool we know of to detect other beings was now at hand in the form of sensitive radio receivers attached to large antennas. Soon, many other astronomers began looking for extraterrestrials' signals.

To make sure the conversation with E.T. could be two-way, messages began to be sent from Earth to other civilizations. NASA spacecraft carried pictures and sound recordings to entertain and educate any alien who may find them in the far future. A powerful radio signal was transmitted to a distant star cluster, using a mathematical code to tell them about our solar system and the molecules of earthly life.

Sometimes, the "phone" rang, and E.T. seemed to be there. In 1967, astronomers found radio signals from the sky unlike any that had ever been heard before. The sharp signals were precisely spaced pulses, and so bizarre that the sources were nicknamed "Little Green Men." But careful study showed that they were something else: radio waves from a strange type of collapsed star.

NASA tried to get into the SETI business by proposing the most powerful search for extraterrestrial intelligence in history. Unfortunately, an influential senator opposed this program, causing a lot of heartache among the scientists of SETI.

Fortunately, not every nation had such powerful opponents to SETI, and the search was carried out sporadically on a smaller scale by Canada, France, West Germany, Holland, Australia and the Soviet Union. And while NASA struggled to get its program approved, a new private organization, The Planetary Society, went to Harvard and helped one of their professors build an advanced SETI project influenced by the NASA design.

Meanwhile, the possibility of interstellar travel slowly worked its way from fiction into the pages of scientific journals. Ingenious ideas were proposed to overcome the enormous distances between stars. The idea of such travel, together with the possibility of extraterrestrial life, gave a hint of respectability to the study of "flying saucers," although most scientists remained highly skeptical.

Finally, after so many years of resistance to the ideas of SETI, the scientific community came to accept the search as something worth doing. But the skeptics are still with us. Some scientists claim that the evolution of life is so complex, the chances of it happening elsewhere are infinitesimal. Some claim that Earth is incredibly lucky to be exactly where it is—not too close to the Sun and not too far—so that the chances of another world escaping a permanent ice age or a perpetual steam bath are almost nil.

We *are* alone, say these voices.

But as of now, with as much objectivity as this author can muster, it looks as if the evidence is on the side of E.T. It's a problem full

NASA's 210-foot diameter Goldstone antenna in the southern California desert. This is used mainly to pick up signals from spacecraft, but one such NASA antenna has been used for SETI in the past, and others will probably be used this way in the future.

of uncertainties, but it seems to me that life very likely exists elsewhere. Some of it is probably intelligent. The chances are, there are signals just waiting for us to pick up. This book shows not only why many of us are optimistic about the search but also why others are pessimistic.

Other civilizations, if they exist, may offer us the fruits of billions of years of wisdom.

And never before has there been such a good time to look.

Martians and Their Friends

ALIENS IN HISTORY

Can there be a place on earth where things are upside down, where the trees grow downwards, and the rain, hail, and snow fall upward? The mad idea that the earth is round is the cause of . . . [this] imbecile legend."
Lactantius Firmianus, A.D. 303-4

IN ROME, on February 17, 1600, a man was taken by the Inquisition to a place of lawful execution. The man: Giordano Bruno. His crime: heresy. Among his beliefs: the Earth is not the center of the universe; there is an infinite number of other worlds; life exists on them . . . The sentence: burning at the stake.

The sentence was carried out. He perished, but his ideas lived on, and helped us begin to understand the universe we live in, and the possibility that others live there too. Bruno was one of the extraordinary people who have led us in our exploration of life and the cosmos, laying the foundations for our present ideas about the possibility of extraterrestrial beings.

Who was Bruno? He helped plant the seeds for SETI by beginning to understand the possibility of other worlds. He was an admirer of the great Polish astronomer, Nicolaus Copernicus, the first man in modern times to seriously propose that the Earth was *not* the center of the universe.

Copernicus dared to challenge the view of most ancient Greeks that the Sun and planets revolved around the Earth. For thousands of years, this obvious "truth" had been believed by almost every human being. (One Greek, Aristarchus of Samos, had suggested that the Earth revolved around the Sun, but such nonsense was dismissed as heresy and the comforting, Earth-centered teachings of

Aristotle were not seriously challenged until Copernicus. Other Greeks and Romans such as Leucippus, Democritus, Epicurus and Lucretius accepted the idea that there were many worlds in the universe, but their ideas died out until Copernicus.)

Humanity enjoyed being the center of the universe until the fifteenth century, when Copernicus arrived on the scene. He was a university student at the tender age of nineteen when Columbus discovered the New World and forced Europe to accept that the Earth was not flat, a shock similar to that which faced the human race five centuries later, when they woke up and found there was an artificial satellite called *Sputnik* orbiting this planet.

Columbus' voyages probably helped stimulate Copernicus into rethinking the Earth's place in the universe. The ancients knew of the planets Mercury, Venus, Mars, Jupiter and Saturn; they were "wandering" stars that moved among the ordinary, "fixed" stars from month to month and year to year.

Copernicus studied the Greeks, especially the great astronomer Ptolemy, who had worked out in elaborate detail the mathematics needed to explain the complicated motion of the planets in the sky. Copernicus thought the whole picture was too messy. If you just let the planets move in circles around the Sun, with each planet moving at a different speed, the picture becomes enormously simpler. And he also achieved a major step toward the ultimate goal of science: to explain complicated phenomena with as few principles as possible.

Copernicus knew perfectly well that his ideas were not going to be met with open arms by the Church. The Church's view of the universe was based on faith, while science, which Copernicus was helping to create, was based on *not* accepting anything on faith. So Copernicus kept much of his work secret until the end of his life. It wasn't until his last year, 1543, that he let his masterpiece, *On the Revolution of Heavenly Bodies*, be published.

Copernicus was an official of the Church known as a canon. This canon fired the intellectual shot heard 'round the world. Giordano Bruno was born just five years after the Polish astronomer died, and became one of those who heard the shot. A priest and philosopher, Bruno felt that most people he came into contact with in the religious and academic worlds were idiots (which was smart) and he told them so (which wasn't). As a result, he spent his life being kicked out of Italy, France, Switzerland, England and Germany.

One of his works, *On the Infinite Universe and Worlds*, shows us how farseeing he was. "Thus there is not merely one world, one Earth, one Sun, but as many worlds as we see bright lights around us. . . . " In this, he anticipated the thinking of many SETI scientists today.

In the same work, he has a character, Burchio, ask, "Then the other worlds are inhabited like our own?"

The character representing Bruno replies, "If not exactly as our own, and if not more nobly, at least no less inhabited and no less nobly. For it is impossible that a rational being fairly vigilant, can imagine that these innumerable worlds . . . should be destitute of similar and even superior inhabitants; for all are either themselves suns or the sun doth diffuse to them no less than to us those most divine and fertilizing rays, which convince us of the joy that reigns at their source and origin; and bring fortune to those stationed around who thus participate in the diffused quality."

You would think that, with a history like his, Bruno would have been wise enough to stay away from his homeland, where the Inquisition was flourishing, but nostalgia won over wisdom and he trotted back to Italy.

He shouldn't have. A "friend" betrayed him to the Inquisition. Bruno spent six years in prison, trying to prove that he was a philosopher, not a heretic, but the Inquisition disagreed. One of the Church's beliefs, handed down from Aristotle, was that the universe was not infinite. According to the reasoning of the philosophers and theologians of the age, only God could be infinite. Nothing material could be infinite—it would challenge God's special place.

Here are Bruno's own words to the Inquisition:

> I hold the universe to be infinite, as being the effect of infinite divine power and goodness, of which any finite world would have been unworthy. Hence I have declared infinite worlds to exist beside this our Earth; I hold with Pythagoras that the Earth is a star like all the others which are infinite, and that all these numberless worlds are a whole in infinite space, which is the true universe.

Bruno knew what awaited him, much as Socrates accepted the cup of hemlock. As the philosopher Giorgio de Santillana once noted:

> In the act of sacrifice, Giordano Bruno knew he was fully resolving the mistakes of his life, the conflicts of his nature, the insufficiency of his intellect, the inadequacy of his equipment. As a man, he knew he had been a failure; but the philosopher in death was transcending that sense of metaphysical failure which dogs the life of the scholarly philosopher however great, and makes him look for a substitute bliss in the fond hope of being raised some time from the purgatory of the footnote to the paradise of the paragraph.

... But we like to imagine that he would not have disdained—for he too was capable of lightheartedness—the words of a contemporary of ours, Lauro de Bosis, written before meeting a similar fate: "There is a message which has to be delivered; whether I live or die has little importance. But if I die, there is a chance that it may go registered and special delivery."

Galileo vs. the Establishment

Bruno's fellow Italian, the brilliant astronomer Galileo Galilei, took Copernicus one step further. Galileo had learned of the invention of the telescope, formed by putting two lenses together in a tube. Where others saw its military possibilities—bringing the enemy closer, making it a "spyglass"—he turned it toward the sky and was astonished at what he saw. The Moon had mountains much like Earth's, and seemed even to have oceans. Jupiter had four moons orbiting around it, making it a scale model of the Copernican solar system, with Jupiter acting like the Sun, and the moons like planets. Venus went through phases just like the Moon. Thus Galileo became the first human being to really see the universe.

There was no way the ancient Earth-centered universe could produce such phenomena. So what did the Establishment do when Galileo spoke up? Naturally, it attacked him.

The Church had almost everything going for it: faith, tradition, inertia, the majority opinion and *power*. All Galileo had was facts.

He lost.

Galileo, after being shown the instruments of torture used by the Inquisition to persuade suspected heretics to be good Christians, recanted his Copernican ideas and lived under house arrest for the remainder of his life.

But those ideas could not be squelched, and when Martin Luther and his followers became influential (although they too opposed the new ideas), at least some freedom of inquiry was permitted in the regions of Europe where they held sway. Nevertheless, even Luther's attitude showed how much he had in common with the church he was attacking: "People give ear to an upstart astrologer [Copernicus] who strove to show that the earth revolves, not the heavens or the firmament, the sun and the moon. . . . This fool wishes to reverse the entire science of astronomy."

The ideas of Copernicus did take seed and gradually sprouted into the brilliant flowers of Newton, Halley, Leibnitz and a horde of other scientists who threw tradition onto the compost pile and brought open-minded experimentation into the the laboratory. Science flourished. It is fitting that the very year Galileo died, 1642, Isaac Newton was born.

In terms of making the theory of astronomy (astrophysics) possible, Sir Isaac Newton probably did more than any other person. Aside from discovering the principle of universal gravitation, in which he developed the three laws of motion (basic physics), Newton devised the mathematics necessary for their computation (calculus). He also revolutionized the field of optics, as in his proof that white light was composed of the colors of the rainbow.

Microscopic Galileo

In Newton's time, there was a man who did for living organisms what Galileo had done for the sky. This man was Anton van Leeuwenhoek, a Dutchman who used the microscope to discover a world as surprising and unusual as the external universe had become. His discoveries were some of the first steps in the understanding of life, toward biology as we know it today. The new science and its sister, astronomy, would blossom in this era, eventually entwining in the pursuit of SETI.

This Galileo of the microscopic cosmos was no conventional scholar. He had not labored in universities studying Greek and Latin, as had most scientists of his day. He was an uneducated shopkeeper, who used crude, expensive, commercially available magnifying glasses to examine cloth. But he was also as thrifty as the proverbial Scotsman. When a lens broke, to save money, he ground one himself. He found he could make a better lens than he had bought, and saw more details in the fiber of the cloth than he had ever seen before.

He began making lenses more and more powerful, until they were better than any made previously. His curiosity drove him to examine familiar objects to see what no human had seen before: the heretofore invisible, intricate structures of skin, hair, plants, wood, bugs, even the eye of an ox.

And a drop of stagnant lake water. Could there have ever been anything less promising than a drop of lake water? But it was there that van Leeuwenhoek made a staggering discovery. For in that drop of water he saw living things, "little beasties." An incredible variety of tiny creatures with tails and horns and legs swam, cavorted, battled, lived and died. He had found an alien world in a droplet.

The American bacteriologist Paul de Kruif wrote, "Beasts these

"I GUESS YOU CAN SAY THAT SINCE LEEUWENHOEK OUR FAMILY HAS BEEN IN SHOW BUSINESS."

were of a kind that ravaged and annihilated whole races of men ten million times larger than they were themselves. Beings these were, more terrible than fire-spitting dragons or hydra-headed monsters. They were silent assassins that murdered babes in warm cradles and kings in sheltered palaces."

Before van Leeuwenhoek, the smallest creatures known were cheese mites, tiny spiderlike critters just barely large enough to see with the naked eye. He wrote, "I would put the proportion thus: as the size of a small animalcule in the water is to that of a mite, so is the size of a honey bee to that of a horse." Thanks to van Leeuwenhoek, people now began to dimly realize that organisms like us were made up of tiny, invisible cells. One day that knowledge would give us powerful clues to the origin of life itself.

Prior to this Dutch shopkeeper's discoveries, life was generally thought to be a mysterious force somehow infecting certain types of matter. Life was supposed to form spontaneously: fleas grew out of dust. Van Leeuwenhoek, however, ignored what everybody knew was so, and proved them wrong. With his microscopes, he found that fleas grew from eggs just like any other insect. This was the first major step in combatting the theory of "vitalism" which held that mystical forces had to be invoked to explain life. It took centuries, but eventually most scientists came to realize that life can apparently be explained by the laws of physics and chemistry. Van Leeuwenhoek pointed the way.

Happily, while so many pioneers have been persecuted by their contemporaries, van Leeuwenhoek was welcomed. The Royal Society of London made him a member, though some grumbled at his uneducated background. He was visited by the King and Queen of England, the German Emperor, and even the Russian Czar, Peter the Great.

For *two thousand years* the magnifying glass had existed. Lenses of glass and quartz were found in the ruins of ancient Pompeii and Nineveh. For two thousand years, anyone could have done what Galileo and van Leeuwenhoek did. We could have begun to understand the universe and the cells of life in the time of Christ. Modern medicine might have developed centuries earlier. The diseases that plague us today might long since have become mere curiosities of medical history.

But nobody did it. How many discoveries just as great could the reader make if he or she were to buy some little electronic gizmo at Radio Shack, or a piece of hardware from Sears and Roebuck, and use it in a way no one has ever done before?

The possibilities are unlimited.

People on Jupiter?

The ideas of Copernicus and van Leeuwenhoek and their colleagues inspired others to start thinking about the possibility of life on other worlds. After all, if Earth is just another planet, then might not Jupiter and Mars and the others be inhabited?

It became popular to speculate about creatures on other worlds. For instance, in his work *Cosmotheoros*, the great seventeenth-century Dutch astronomer, Christian Huygens, envisioned plants, animals and intelligent beings living on all the other worlds of the solar system. Published after his death in 1695, a decade after Newton's masterpiece, the *Principia Mathematica*, it speculated on how similar they might be to us:

> . . . it might nevertheless be reasonably doubted, whether the Senses of the Planetary Inhabitants are much different from ours. . . . Men . . . reap Pleasure as well as Profit, as from the Taste in delicious Meats; from the Smell in Flowers and Perfumes; from the Sight in the Contemplation of beauteous Shapes and Colours; from the hearing in the Sweetness and Harmony of Sounds. . . . Since it is thus, I think 'tis but reasonable to allow the Inhabitants of the Planets these same Advantages that we have from them.

He even speculated that alien sailors might sail planetary seas:

> Especially considering the great Advantages *Jupiter* and *Saturn* have for Sailing, in having so many Moons to direct their Course, by whose Guidance they may attain easily to the Knowledge that we are not Masters of, of the Longitude of Places. And what a Multitude of other Things follow from this Allowance? If they have Ships, they must have Sails and Anchors, Ropes, Pullies, and Rudders, which are of particular Use in directing a Ship's Course against the Wind, and the Sailing different Ways with the same Gale. And perhaps they may not be without the Use of the Compass too, for the magnetical Matter, which continually passes thro' the Pores of our Earth, is of such a Nature, that it's very probable the Planets have something like it.

In at least one way, he was correct: we now know that many planets do indeed have magnetic fields "passing thro' their Pores."

The Evolution of Darwin

Physics and chemistry grew in the next century until the scientific world was ready to assimilate this knowledge into the emerging science of biology. The very word *biology*, the science of life, wasn't

Dueling dinosaurs that "made hideous the waters of Central Europe," according to a nineteenth-century writer: Ichthyosaurus (L.) and plesiosaurus.

A giant land reptile, the hadrosaurus, from a skeleton found in the nineteenth century.

invented until 1800, mostly because scientists had been content simply to catalog living things and their parts, and to limit their thinking about the fundamental nature of life to mystical philosophizing. The nineteenth century saw the systematic use of physics and chemistry on this problem.

Nine years after the birth of biology, a child was born who would give the new word profound meaning. The child was Charles Darwin, and his genius finally wove the entire tapestry of life into a single, beautiful, and, above all, *logical* picture no longer dependent on superstition.

Darwin did not work in a vacuum. Astronomy, in the wake of Copernicus and Newton, had made scientists aware that the universe operated according to knowable laws, and made people wonder whether the seeming magic of life itself might obey such laws. Scientists before Darwin had speculated about the possibility that life might have evolved from simple things to complex creatures. Astonishingly, the French naturalist Georges Buffon anticipated Darwin when he wrote in 1783, "that man and ape have a common origin; that, in fact, all the families among plants as well as animals, have come from a common stock." But it just remained one of many speculations until Darwin came along.

Geology provided some of the clues Darwin needed. The study of rocks, especially fossils, showed that the Earth had a history. An incredible zoo of living things that no longer existed, from shellfish to dinosaurs, was preserved in the fossil record Darwin studied. In 1822, the year Darwin became a teenager, dinosaur remains were first discovered. Fossil plants had been known since antiquity, but on this occasion, English fossil hunters found bones of a twenty-foot-long monster.

The discovery came about because a doctor had a hobby of fossil hunting. Gideon Algernon Mantell was roaming with his wife through a British forest when Mrs. Mantell found the first bone. After first being misidentified as the teeth of a rhinoceros and the bones of a hippopotamus, the skeleton was finally recognized as that of a lizard far larger than any now living. They christened the creature *iguanodon*, because of its iguana-lizard teeth.

Mantell was hooked. He became obsessed with dinosaurs. So crazy was he about fossils that he eventually filled their house with them, causing his wife to leave him. Such are the hazards of science.

Fossils indicated that something was amiss in the Biblical record. In 1654, Archbishop James Ussher had counted all the "begats" in the Bible and concluded that "the world was created on 22nd October, 4004 B.C. at 6 o'clock in the evening." Pretty precise, eh? But science had already found that the Earth had to be at least millions of years old to explain the geological record. Maybe the Archbishop wasn't so precise after all.

Darwin grew up to be a naturalist who studied life in the placid English countryside until he had a chance to join a ship with the unpromising name of H.M.S. *Beagle*. In 1831, Darwin embarked upon a grueling scientific cruise on the *Beagle*, one that was to last until 1836, taking him along the coast of South America, to the Galápagos Islands in the Pacific, on to New Zealand and Australia and into the history books as the Copernicus of biology.

Where Copernicus had insulted our human pride by suggesting that the universe did not revolve around us, Darwin gave us even more humiliating news: we were not even special creatures on this planet. All life was descended from a common heritage.

He came to this conclusion by painstakingly studying the plants and animals he came across on his long journey. He saw how insects adapted so they would blend in with their surroundings and thereby be less likely to be eaten. He saw that the same species of birds in the Galápagos were different on each successive island, each subtly adapted to the foliage and food available. The constant flood of new species, new variations on old species, and new environments stimulated his powerful brain. "My mind," said he, "seems to have become a kind of machine for grinding general laws out of large collections of facts."

Darwin came to realize that everything was connected, all lifeforms interrelated. He realized that life on Earth had started out in some very simple form such as a cell. It seemed life was a constant battle for food: "eat or be eaten" was the First Commandment. Those critters who could move a little faster or were a little stronger or a little smarter were more likely to be among the happy

"The picture's pretty bleak, gentlemen. . . . The world's climates are changing, the mammals are taking over, and we all have a brain about the size of a walnut."

eaters than among the unfortunate eaten. The winners would reproduce more often than the losers.

And occasionally, nature goofs. It produces mutations, departures from the blueprint of the species: two-headed calves, Siamese twins. Most of these mutations are useless and die off before reproducing. But occasionally one might be useful. A fish capable of breathing air might survive being tossed ashore. The offspring of a normally hairless creature born with fur might thrive in a cold climate while its naked brothers freeze to death. An ape with highly mobile thumbs might be better able to use tools than one with five ordinary fingers on each hand.

In this way, generations of lifeforms adapt to different environments and complex organisms build up from simpler ancestors. Thus, humans were not just remarkably similar to apes, but descended from a common ancestor. Indeed, our lineage could be traced back to worms and beyond. The worm takes food in at one end, digests it in an internal tube and excretes it at the other end (a crude model of a human, but with remarkable similarities in the many different organs that make both creatures work).

At first, Darwin's theory was hardly noticed. In 1858, after he presented his theory to a scientific society, its president remarked that the year "has not, indeed, been marked by any of those striking discoveries which at once revolutionise, so to speak, the department of science in which they occur." Later, and more accurately, Ashley Montagu remarked that "next to the Bible no work has been quite as influential, in virtually every aspect of human thought."

Darwin finally published his ideas in the classic *Origin of Species*, which quickly became a bestseller. The Darwinian word spread and the mystical view of the origin of life began to sink like the *Titanic*, though many survivors struggled in their lifeboats to keep the old views afloat. Surprisingly, many scientists were in those lifeboats. The leading American zoologist, Louis Agassiz, called evolution "a travesty of facts."

Imagine the impact on a stuffy Victorian Englishman. To the upper-class British, as with most other societies of the day, your ancestors were practically everything. Kings, Queens and a whole frosting of lesser nobles determined their pecking order by the sexual encounters of their ancestors, much like horse breeders (though with less planning). Your grandfather might have been a psychopath, a cutthroat and a child molester, but if he was of royal blood, then he was obviously superior to the chap in the street, and, more to the point, so were you.

Now along comes Darwin. If your relatives are so bloody impor-

A Victorian gent shows off his fossil sigillaria tree-trunk.

tant in the scheme of things, what do you think about being cousin to *apes*, grandchild of *worms*? Egad, Quimby, the smelling salts! You can see why an Englishwoman remarked, "I certainly hope we are not descended from the apes; but if it should be true, I pray that it does not become common knowledge." She should have listened to the American biologist J. M. Tyler, who said, "Even if we are descended from worms they were glorious worms."

But it was even worse than that. For more than two thousand years, the Bible had been the ultimate Word on the origin of life. The Book of Genesis said that God created Earth and all its life in six days, and on the seventh, He rested. Even though Darwin was a religious man, he did not think a book written for simple shepherds two thousand years ago should be taken as the literal, nuts-and-bolts story of the history of life. In *Origin of Species*, he wrote, "I see no good reasons why the views given in this volume should shock the religious feelings of anyone."

But the most literal-minded interpreters of the Bible claimed that Darwin was blasphemous. In 1860, a landmark event occurred in the battle between evolution and religion—a battle that is still being fought today. While Abraham Lincoln was running for President of the United States, over in England, at Oxford University, biologist Thomas Henry Huxley debated Bishop Samuel Wilberforce. Darwin was ill and could not attend, so Huxley served as his champion before the hostile audience. The Bishop, known as Soapy Sam, basked in the obvious approval of the audience, and mocked "the monkey theory" at length. After he had spoken eloquently and wittily, he turned to Huxley and said, "Was it through his grandfather or his grandmother that he claimed his descent from a monkey?"

Huxley was delighted with this opening into the shallow mind of the Bishop. He whispered to a friend, "The Lord hath delivered him into mine hands."

The Bishop sat down to tumultuous applause and Huxley took his place. Piece by piece, Huxley dissected the Bishop's arguments and exposed Soapy Sam's ignorance of the observations that had been so painstakingly gathered by Darwin. Fact piled upon fact and each of the Bishop's arguments was answered. The audience was unable to resist the impact of combined rhetorical skill and logic. "A man," said Huxley, "has no reason to be ashamed of having an ape for a grandfather. If there were an ancestor whom I would feel shame in recalling, it would rather be a *man*, a man . . . who . . . plunges into scientific questions with which he had no real acquaintance, only to obscure them by an aimless rhetoric, and distract . . .

The hairy mammoth, which added excitement to the lives of our caveman ancestors, from a nineteenth-century drawing. One cave-person drew a picture of these beasts on a piece of ivory.

"Was it from your grandfather or your grandmother that you claim descent from a monkey?"

his hearers from the real point . . . by . . . digressions and skilled appeals to religious prejudice."

The audience roared its approval. A lady fainted. For at least one day, science triumphed.

The theory did not become instantly accepted, however. British scientist and clergyman Francis Orpen Morris, for example, wrote, "If the whole of the English language could be condensed into one word, it would not suffice to express the utter contempt those invite who are so deluded as to be disciples of such an imposture as Darwinism." British geologist Adam Sedgwick said, "I laughed . . . till my sides were sore." In this century, William Jennings Bryan said, "All the ills from which America suffers can be traced back to the teaching of evolution. It would be better to destroy every other book ever written, and save just the first three verses of Genesis."

But evolution did soon become accepted by open-minded people, and the theory influenced people far removed from science. Socialists loved its radical challenge to the Establishment. Karl Marx, who saw human history as a series of changing politico-economic systems rather like Darwin's survival of the fittest, wanted to dedicate *Das Kapital* to Darwin. The honor was declined. Marx must have been appalled when many politicians and social philosophers eventually embraced it as a justification for free-market economics, calling it "Social Darwinism." Ironically, in the twentieth century, Marx's followers under Stalin would suppress Darwinism in favor of a discredited theory descended from an earlier French scientist, Lamarck.

Incidentally, another Englishman, Alfred Wallace, had the same basic idea as Darwin at roughly the same time, while suffering from fever in the Brazilian jungle. When Darwin heard of that, he knew he had to publish his theory quickly or he would suffer the greatest hell known to the modern scientist: the entry in the record books would list him as the *second* guy who made the discovery. Now that the historical dust has settled, we can say that they both deserve credit, though Darwin was first and did far more to prove the concept of evolution. For that reason, it is sometimes referred to today as Darwin-Wallace evolution.

Victorian SETI

The ever-widening horizons of astronomy and the ever-more scientific thinking about the history of life on Earth stimulated ideas about the possibility of life on other worlds, the central idea of SETI.

A Lunarian, from an Italian translation of the Moon Hoax of 1835.

If life had happened here, reasoned many nineteenth-century thinkers, why couldn't it have happened elsewhere? And if it were up there in the sky somewhere, couldn't a civilization that had produced such high technology as the steam engine and electricity communicate with them?

This type of thinking became increasingly common as traditional dogmatic religion lost its grip and science gained credibility. If you were beginning to doubt the literal truth of the Bible, it was comforting to think there might be friends on other worlds.

Thus was born one of the least-known scientific ventures of the nineteenth century: the attempt to signal other civilizations in space.

One astronomer claimed to have seen great fortifications and other buildings on the Moon. Then, in the 1820s, the great German mathematician, Karl Friedrich Gauss, proposed communicating to the lunar inhabitants by planting pine trees over a vast area of Siberia. The trees would be laid out in the shape of squares on the sides of a right triangle. Any Lunarians with a telescope would then be able to see that we Earthlings had mastered the theorem of Pythagoras, that the sum of the squares of the legs of a right triangle equals the square of the hypotenuse. This would prove to them that not only were there living things on Earth, but mathematically speaking, we were housebroken.

The public was fascinated with these ideas and entranced by the march of science. People were ready for anything when, in 1835, the *New York Sun* reported on the expedition of a major British astronomer, Sir John Herschel. He was the son of a great German-turned-British astronomer, Sir William Herschel, who had, in addition to discovering the planet Uranus, speculated enthusiastically about the possibility of life on the Moon.

Sir John was trying out a new telescope in South Africa, where he could see the southern part of the sky, which was inaccessible to those living in the Northern Hemisphere. The *Sun*'s reporter wrote a series of articles detailing the astronomer's discoveries. He wrote that Herschel had seen the lunar inhabitants through his telescope:

> [They] averaged four feet in height, were covered, except on the face, with short and glossy copper-colored hair, and had wings composed of thin membranes. . . . Our further observation of the habits of these creatures, who were of both sexes, led to results so very remarkable, that I prefer they should first be laid before the public in Dr. Herschel's own work . . . they are doubtless innocent and happy creatures, notwithstanding some of their amusements would but ill comport with our terrestrial notions of decorum.

The articles came complete with drawings of creatures much like apes. There was only one thing wrong with them. They were fakes. Herschel never made any such reports. A writer on the *Sun* had written these stories as a satire, but readers took them seriously; about half the people of New York City believed every word. It is a fascinating measure of human gullibility that even after the writer admitted his fakery, many people continued to believe in it.

This journalistic prank became known as the Moon Hoax. In these days of widespread belief in flying saucers and Bermuda Triangles, we would do well to bear it in mind.

The Martians

By 1840, the Viennese astronomer Joseph von Littrow, evidently not an enthusiast of the Siberian climate which Gauss had preferred for his Pythagorean triangle, suggested building a twenty-mile wide ditch in the Sahara desert. He planned to fill it with kerosene and light it. At night, the sight would attract the attention of our celestial neighbors. (In this century, something like that has been done for other reasons. Arabs burn excess natural gas from some of their oil wells. These have been seen from space, but only by human astronauts as far as is known.)

One person wanted to build huge mirrors across Europe.

None of these grandiose projects was built, but a man came along who did more for this game than any other. He was the one who really put Mars on the map: Percival Lowell. In his early school years, he was fascinated with the subject of astronomy. He was a Boston Brahmin, born to a prominent family, and like most such gentlemen, was consigned to a Harvard education. He focused his studies on mathematics, a subject a true Boston gentleman would have preferred to ignore once it ascended beyond the level needed to understand the stock market.

After Harvard, he roamed the Far East for his father's cotton business, spending a decade in Japan and other exotic locales. When he returned to the United States in 1893, he could no longer contain his secret love of astronomy. He came out of the closet and founded an observatory in Flagstaff, Arizona. It was a town in the wide-open spaces of the wild west, a paradise for gila monsters and astronomers in the days before city lights and smog threatened even such remote places.

His life was changed by the work of the Italian astronomer Giovanni Schiaparelli, who had observed Mars and had seen markings that he called *canali*, using the term of his colleague Pietro Secchi. The word *canali* simply means channels, which can refer to natural

Percival Lowell, the man primarily responsible for the world's fascination with Mars.

The dome at Lowell Observatory, containing Lowell's 24-inch telescope.

grooves in dirt left by water and other processes, but it was inevitably mistranslated and publicized as "canals." And since *canals* are not made by Mother Nature, they must have been built by Martians.

Never has a mistranslation served astronomy so well. Lowell became fascinated with the idea of canals on Mars. This should not be too surprising, since the nineteenth century might well be called the Century of Canal Consciousness. In 1855, the year of Lowell's birth, Ferdinand de Lesseps was given the Suez Canal concession in Egypt. Fourteen years later, this canal, the greatest in history, was opened by Empress Eugenie. When Lowell was twenty-two, Schiaparelli published his observations of the *canali* on Mars. Two years later, de Lesseps formed the Panama Canal Co. to build the next great waterway (although that effort went bankrupt a decade later).

When Lowell built his observatory, his sole purpose was to study the advanced civilization of Mars. Over the years, he bought larger and larger telescopes and conducted expeditions to observe from better locations. His trip to the Chilean Andes resulted in the first high-quality photographs of Mars. He was as obsessed with Mars as Gideon Algernon Mantell had been with dinosaurs, and he became the world's greatest expert on the planet.

There were seasonal changes in the colors of Mars. In the Martian summer, dark areas would form near the canals. Clearly, water was being brought in to supply vast farms. To produce a planet-wide network of canals was beyond the power of nineteenth-century technology, so clearly the Martians had to be more advanced than Victorians.

At one time, Lowell even claimed to have observed canals on Venus. We now know that Venus has an atmosphere much thicker than ours and is perpetually overcast, so the Venusians would have had to build their canals above the clouds. When none of Lowell's contemporaries could find the Venusian canals, astronomers returned to studying those on Mars.

Lowell did not just publish research papers in learned journals the way most scientists do. He wrote books in plain English for the public: books like *Mars* (1895), *Mars and Its Canals* (1906), *Mars as the Abode of Life* (1908). The public ate them up. Some astronomers thought they saw lights on Mars. Before long, people were sure a Martian civilization existed, and the idea of communicating with it became irresistible. As Carl Sagan once observed, there was no doubt that intelligence was involved. The only question was, on which end of the telescope was the intelligence?

So popular did the idea of life on Mars become that a contest was eventually held to reward the first discoverer of intelligent life

Lowell's telescope with its 24-inch-diameter lens, in Flagstaff, Arizona, with which Percival Lowell studied Mars.

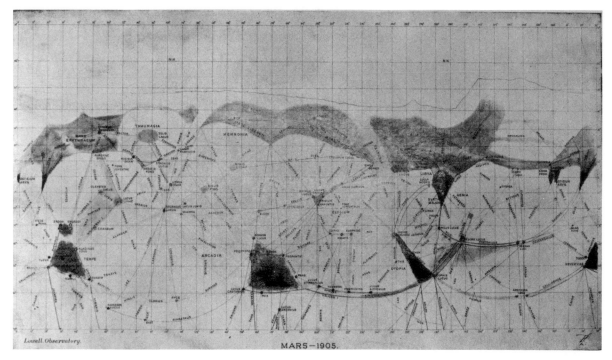

Percival Lowell's 1905 Mars map with the "canals" carefully labeled. We now know that the markings are a combination of dark Martian features and optical illusions.

beyond Earth, excluding Mars—because that would be too easy.

The American physicist-poet Robert Wood (author of *How to Tell the Birds from the Flowers*) had the idea of placing giant black cloths across white alkali planes. By moving them back and forth with motors, Earth could "wink" at Mars.

The British scientist Sir Francis Galton, a cousin of Darwin in the legal (not just Darwinian) sense, invented a Morselike code to transmit the coordinates of a picture. (This idea would be revived repeatedly in the twentieth century.)

The French astronomer Camille Flammarion wrote in 1892, "The present inhabitation of Mars by a race superior to ours is very probable," and thought there must be a better way of communicating with other worlds. He suspected that in centuries to come, "a new and unexpected discovery" would be invented that did not involve building giant machines.

Often, when a scientist forecasts a great discovery far in the future, he is apt to stumble over someone who is making the same discovery at that very moment. Such was the case then.

A German physicist, Heinrich Hertz, had just recently discovered radio waves, which had been predicted theoretically by the Einstein of nineteenth-century physics, a Scot named James Clerk Maxwell. Hertz' name today is honored as the unit of frequency measurement.

Had Flammarion given their work some thought, he would have realized that the answer to the Martian communication problem was basically solved, and just needed some practical work. He was not alone in failing to see the promise of the new "Hertzian waves."

The man who solved the practical problems was an Italian inventor, Guglielmo Marconi. Hertz could initially transmit signals over a distance of only a few yards—you might as well shout. But by 1895, Marconi was able to transmit code a mile. Radio began to look promising. By the turn of the century, he could transmit across two hundred miles, and he began to dream of sending a signal across the Atlantic. Some scientists pooh-poohed his proposal because—hadn't you heard, Marconi? The Earth is round! Radio signals, like their cousins the rays of light, travel in straight lines. Any signal will obviously go off into space or be absorbed by the Earth.

Any experimenter knows you should never trust a theoretician. So Marconi ignored the skeptics, and on a memorable day, December 12, 1901, he transmitted a signal between Cornwall, England, and Newfoundland, Canada, an incredible distance of 2,000 miles. Only later was it realized that the Sun's ultraviolet light ionizes the upper atmosphere, tearing electrons out of atoms, which makes the upper atmosphere (the ionosphere) electrically conductive, like a metal blanket around the Earth. Marconi's signals bounced off the ionosphere and landed in Canada. And that's how the new century saw a radically new means of communication capable of crossing the entire globe.

For the first time, the ocean could be crossed at the speed of light (the speed of radio waves). Of course, it was information and not people that did the traveling, but that was a powerful hint of how messages could be carried across interplanetary space, if they could somehow get through that ionosphere. This was to become the major cornerstone of modern SETI.

American rocket pioneer Robert Goddard thought this ionospheric barrier was a great excuse to send rockets out there, since you couldn't send radio signals. But after the First World War, it was realized that you could indeed penetrate the atmosphere by radio. You just had to use higher frequencies than were used in transatlantic radio. Signals can pass through the ionosphere if their frequency is higher than about 10 MHz (megahertz, a million cycles per second).

For a while in the 1920s, Marconi actually thought he had detected radio signals from Mars. They were probably a strange phenomenon called whistlers, radio waves produced by lightning thousands of miles away. These signals travel into space but are sometimes trapped by the magnetic field of the Earth. Different frequencies travel at slightly different speeds, so that when they re-enter the atmosphere, the high-frequency part may arrive slightly before the low-frequency part, making them sound like God slowly whistling. Very eerie.

Electrical whiz Charles Steinmetz then said, "If the United States should go into the effort to send messages to Mars with the same degree of intensity and thoroughness with which we went into the war, it is not at all improbable that the plan would succeed."

When Mars was near the Earth in 1924, a friend of Lowell's convinced the U.S. Government to listen for Martian messages. Strange signals were heard, and one pair of researchers recorded them on film. They seemed to form a human face. They turned out to be noise from the increasingly popular radio transmitters of Earth. This was a good lesson for those who, even today, readily see signs of extraterrestrials on the Moon, Mars and in Egyptian pyramids.

Around this time, the death of Lowell, failure to detect the Martians and astronomers' growing suspicion that the canals were really optical illusions deflated scientific interest in the possibility of other civilizations. It was up to the science-fiction writers to take over.

Bug-Eyed Monsters and Little Green Men

ALIENS IN SCIENCE FICTION

But who shall dwell in these worlds if they be inhabited? . . . Are we or they Lords of the World?
Johannes Kepler

"Gort, Klaatu berada nikto!"

These are the words from the classic 1951 film *The Day the Earth Stood Still*, used by Patricia Neal to command the menacing alien robot, Gort, to stop. She does this in the nick of time, as the alien emissary, Klaatu (Michael Rennie), lies dying, the victim of a gunshot. (These alien words are immortalized on the wall of a men's room at M.I.T. I have sometimes fantasized that, one day, Earth is invaded by hostile aliens, and I save the world, because I'm the only one who knows in which men's room are written the words to stop the invaders.)

The alien came here to prevent Earth from extending its violence into the galaxy, a theme that resurfaces in real-world SETI when scientists calculate the odds of detecting other civilizations. Violent ones probably destroy themselves before long, a sobering thought that unexpectedly appears in the equations of SETI.

The Day the Earth Stood Still is a beautiful example of the twentieth century's expression of a type of literature born in the nineteenth century: science fiction. The imagination of the writer was stimulated by that of the scientist, and vice versa. Alien creatures which had since antiquity been the province of such mythmakers as Homer were welcomed into science following Copernicus. Since science fiction has produced or influenced many of the ideas behind

31

SETI, it is worth dallying a bit among the fictional aliens to see why this is so. Readers impatient for hard facts should skip this chapter.

In olden days, the yarn spinner could simply invent monsters with the heads of snakes or gods throwing thunderbolts. That sort of thing became disreputable with the rise of Christianity and the downfall of the old gods. But by the nineteenth century, when artists' imaginations once again had free rein, science provided a whole new set of possibilities.

The tremendous growth of science could not be ignored by writers, and they responded by creating a new form of literature. Scientists often dislike the mention of science fiction in discussions of SETI, because they fear that publicity involving such associations will lead politicians to dismiss SETI as mere Buck Rogers nonsense and cut off funds for research.

Yet many scientists got their first intimations of the possibility of extraterrestrial life during a youth spent watching *The Blob*, poring over *Space Patrol* comics and reading *Rocket Ship Galileo*. Without that early influence, many of them today might well be accountants or farmers. The wild speculations of science fiction often led them to wonder seriously about what the universe is actually like beyond the confines and biases of our own planet. As the distinguished British researcher J. B. S. Haldane wrote in *Possible Worlds and Other Papers* (1927), "I have no doubt that in reality the future will be vastly more suprising than anything I can imagine. Now my own suspicion is that the universe is not only queerer than we suppose, but queerer than we can suppose."

So let's look at what happened, from the nineteenth century to the present, as fiction fed upon science, as the intellectual and political climate shaped what writers dreamed up about unearthly worlds, and as the general public began to project its fears and hopes onto other possible civilizations.

Invasion of the Lunatics

The nineteenth century was an age when a cornucopia of scientific and technological marvels was pouring forth: electricity, the steamboat, photography, organic chemistry, radio, the telegraph and telephone, anesthesia, automobiles, dynamite, genetics, X-rays, radio. With a miracle being born seemingly every day, it must surely have appeared that there was nothing science could not do.

Naturally, writers could not resist a future filled with fabulous inventions. Halfway through the century, in 1851, the term *science fiction* was coined to describe this new literature. (The expression did not become commonly used until the next century, as Victorians preferred to call such stories *scientific romances*.) Trips to the Moon had been fantasized about by writers as far back as the Greeks, but in the nineteenth century, logic and science were at last woven into fiction, giving such tales a feeling that they could happen tomorrow.

Although most people think the only authors writing science fiction back then were Jules Verne and H. G. Wells, the extraordinary fact is that most of the greatest American writers of that era and some of their counterparts in Europe wrote science fiction or fantasy. Not until the twentieth century was this field regarded as somehow inferior to mainstream literature. The science-fictional *Who's Who* of that age includes Edgar Allan Poe, Sir Arthur Conan Doyle, Herman Melville, Honoré de Balzac, Nathaniel Hawthorne, Jack London, Ambrose Bierce, Edward Bellamy, Mark Twain, Mary Wollstonecraft Shelley, James Fenimore Cooper, Henry James, Stephen Crane, Washington Irving and even Fyodor Dostoyevsky.

Americans, without the inertia of lengthy traditions, vigorously accepted change in their literature as well as their society, so native writers embraced the scientific fantasy more warmly here than elsewhere.

Washington Irving, for instance, author of *Rip Van Winkle* and *The Legend of Sleepy Hollow*, may not be someone you think of as a science-fiction author, but he was. In one of his Knickerbocker tales, "The Men of the Moon," he picked up the astronomical idea of life up there and envisioned the invasion of Earth by "Lunatics," foreshadowing the interplanetary invaders who would so preoccupy future writers. Published in 1809, the year Darwin was born, this story satirized the taking of America from the Indians, but Irving's powerful imagination laid out some of the ideas about extraterrestrial life that scientists today still debate:

> Many a time . . . have I lain awake whole nights debating in my mind, whether it were most probable we should first discover and civilize the moon, or the moon discover and civilize our globe. Neither would the prodigy of sailing in the air and cruising among the stars be a whit more astonishing and incomprehensible to us than was the European mystery of navigating floating castles, through the world of waters, to the simple natives.
>
> . . . let us suppose that the aerial visitants I have mentioned [were] possessed of vastly superior knowledge to ourselves; that is to say, possessed of superior knowledge in their art of extermination—riding on hyppogriffs—defended with impenetrable armor—armed with concentrated sunbeams, and provided with vast engines. . . .

The captain of the extraterrestrials describes the Earthlings to his leader as differing "in everything from the inhabitants of the moon, inasmuch as they carry their heads upon their shoulders, instead of under their arms—have two eyes instead of one—are utterly destitute of tails, and of a variety of unseemly complexions, particularly of horrible whiteness, instead of pea-green." This may have been the origin of the proverbial Little Green Men.

Irving demonstrated well one of the most important virtues of science fiction, and indeed of SETI: it forces us to see ourselves as others may see us. In *Future Perfect*, the authoritative study of nineteenth-century American science fiction, H. Bruce Franklin says, "Irving thus shows one reason why science fiction was so congenial to America: because it is a nation that originated in conquest by alien beings who voyaged here from another world."

Mother of Frankenstein

The year was 1818. Darwin was a boy of nine; Queen Victoria was not to be born for a year. The first work of true science fiction in the modern sense was published: *Frankenstein*. In the monster's unnatural origin, his unfamiliarity with normal human life and the terror he inspired, it was as if he were from another world.

It had started almost as a joke. Poet Percy Bysshe Shelley and his mistress (and soon-to-be-wife) Mary Wollstonecraft left England and visited with their friend, Lord Byron, in Switzerland. The weather was rainy and miserable, so they spent most of their time indoors. For fun, they read some German ghost stories in French. Inspired, Lord Byron said, "We will each write a ghost story." What a contest: two of the greatest poets of the century pitted against young Mary (she was only nineteen).

One evening, they were talking about the origin of life according to Erasmus Darwin—Charles Darwin's grandfather. Erasmus had his own ideas about evolution before Charles was born. Although Erasmus' theory eventually failed to fit the facts, it inspired Charles toward his own concepts.

It also inspired Mary Shelley.

> Many and long were the conversations between Lord Byron and Shelley to which I was a devout but nearly silent listener. During one of these, various philosophical doctrines were discussed, and among others the nature of the principle of life, and whether there was any probability of its ever being discovered and communicated. They talked of the experiments of Dr. Darwin . . . who preserved a piece of vermicelli in a glass case till by some extraordinary means it began to move with voluntary motion. . . . Perhaps a corpse would be reanimated; galvanism had given token of such things: perhaps the component parts of a creature might be manufactured, brought together, and embued with vital warmth.

The Italian physiologist, Luigi Galvani, had discovered in the previous century that dead frogs sometimes twitched when hung by copper hooks from an iron rail, which his countryman Alessandro Volta discovered was due to electricity generated when different metals touched. Electricity seemed to have something to do with life.

Mary had a sleepless night following the discussion and had a vision which became the basis of *Frankenstein*. The stories Byron and Percy wrote for their little contest have sunk into obscurity, but Mary's creation lumbered onto the pages of literary history.

Science had prepared the public for this gothic tale of the creation of life, and the church had prepared them to hiss at the sacrilegious concept of the scientist daring to create life. This was the now-familiar story of the scientist, Frankenstein, who put together a tall man from a mix-and-match assortment of pieces of bodies. Here are Herr Frankenstein's own words:

> It was on a dreary night of November that I beheld accomplishment of my toils. With an anxiety that almost amounted to agony, I collected the instruments of life around me, that I might infuse a spark of being into the lifeless thing that lay at my feet. It was already one in the morning; the rain pattered dismally against the panes, and my candle was nearly burnt out, when, by the glimmer of the half-

extinguished light, I saw the dull yellow eye of the creature open; it breathed hard, and a convulsive motion agitated its limbs.

How can I describe my emotions at this catastrophe, or how delineate the wretch whom with such infinite pains and care I had endeavoured to form? His limbs were in proportion, and I had selected his features as beautiful. Beautiful! Great God! His yellow skin scarcely covered the work of muscles and arteries beneath; his hair was of a lustrous black, and flowing; his teeth of a pearly whiteness; but these luxuriances only formed a more horrid contrast with his watery eyes, that seemed almost of the same colour as the dun-white sockets in which they were set, his shrivelled complexion and straight black lips . . .

The monster comes to life initially quite innocent—basically, a nice guy, not the evil monster of the movies. But after the scientist refuses his creation's not unreasonable request to build him a mate, the monster goes wild. In the end, the scientist loses his wife, his brother and his own life.

When published, the novel was a huge sensation, and soon became a hit play on the London stage. The story resonated with something deep in the human soul. How else to explain its endurance for more than a century and a half? Its message reverberates today. Who has not worried about the forces science has unleashed? The familiar threat of nuclear weapons is but one monster many have wished could have been left asleep. And genetic research is bringing us closer every day to what Mary Shelley anticipated.

Perhaps there are some things Man was not meant to know.

Aliens Galore

Frankenstein was just the first in an endless line of scientifically conceived monsters lurching across the pages of fiction. Many of his literary descendants evolved into beings on other worlds.

In France, Jules Verne was fascinated by the technology roaring through the era. In 1865, just six years after Darwin's *Origin of Species* came out, Verne published *From the Earth to the Moon*. His story of a rocket launching from Florida and taking men to the Moon was, of course, not to become reality until a hundred and four years later, but in the meantime, it provided the first literary trip to another world by "scientific" means. Coincidentally, in the same year, the French astronomer Camille Flammarion published *Real and Imaginary Worlds*, speculating about life on the planets.

The next year, the British answer to Jules Verne was born: H. G. Wells. In 1895, he published his first novel, *The Time Machine*, which depicted the effects of evolution on the future of the human race

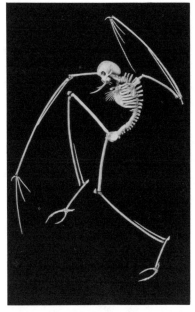

The skeleton of an extraterrestrial creature, the first in a sequence of creatures envisioned by artist/sculptor Joel Hagen, who builds and paints the aliens we may one day find.

as producing two new species: the nasty, ratlike Morlocks, living underground, and the seemingly idyllic Eloi, living above.

Wells soon began applying this evolutionary thinking to another world, and in 1898, when *War of the Worlds* came out, he set a new standard with his Martians. Inspired no doubt by Percival Lowell's speculations about their civilization, his Darwinian thinking encouraged him to see aliens as competitors in the race for survival, a theme that was to dominate science fiction until the 1930s. This attitude also expressed, however subtly, the real-world imperial attitudes toward "inferior races" that climaxed in Nazi Germany.

The Planet of the Pulps

Science fiction owes much of its twentieth-century popularity to the technology of the paper industry. In the 1880s, a method of making cheap paper from wood pulp was developed. "Pulp" magazines soon appeared, cheaper than the "slick" ones that appealed to a wealthier readership. The growth of advertising allowed costs to be reduced even further, and mass distribution brought magazines to virtually every town.

Four years before the turn of the century, publisher Frank A. Munsey converted the pulp magazine *The Argosy* into an all-fiction format that established the formula for numerous competitors. Soon, "the pulps" were publishing all manner of popular fiction: mystery, horror, adventure and the occult as well as fantasy and science fiction.

In 1912 there appeared a serial in one of those pulps which was to inspire generations of scientists and writers. *Under the Moons of Mars* appeared in *All-Story Magazine* under the name of Norman Bean. (The author's real name was Edgar Rice Burroughs, and the same year, in the same magazine, he also created one of the century's most popular characters, Tarzan.) The hero was Confederate Captain John Carter of Virginia, who starts his adventures just after the Civil War by escaping from hostile Apaches in the wilds of Arizona. He enters a cave and is miraculously transported to a world known to its natives as Barsoom, or to humans as Mars.

Under the Moons of Mars was published in book form as *A Princess of Mars*, and it became the first in a long series of novels about that world. Here is Carter's first glimpse of Martians, freshly hatched from eggs:

> . . . the grotesque caricatures which sat blinking in the sunlight were enough to cause me to doubt my sanity. They seemed mostly head, with little scrawny bodies, long necks and six legs, or, as I afterward

Skull: "Endohyrax."

learned, two legs and two arms, with an intermediary pair of limbs which could be used at will either as arms or legs. Their eyes were set at the extreme sides of their heads a trifle above the center and protruded in such a manner that they could be directed either forward or back and also independently of each other, thus permitting this queer animal to look in any direction, or in two directions at once, without the necessity of turning the head.... Against the dark background of their olive skins their tusks stand out in a most striking manner, making these weapons present a singularly formidable appearance....

Later, he meets the lovely Martian Princess, Dejah Thoris, whose alien physiology is, by an extraordinary feat of Martian evolution, sufficiently similar to an Earthly female's to stimulate Carter's hormones:

> ... the sight which met my eyes was that of a slender, girlish figure, similar in every detail to the earthly women of my past life.... Her face was oval and beautiful in the extreme, her every feature was finely chiseled and exquisite, her eyes large and lustrous and her head surmounted by a mass of coal black, waving hair, caught loosely into a strange yet becoming coiffure. Her skin was of a light reddish copper color, against which the crimson glow of her cheeks and the ruby of her beautifully molded lips shone with a strangely enhancing effect.

The reality of evolution is that our present appearance is such a hodge-podge of four billion years of accidents that it is very unlikely that there are little green female aliens appealing enough to be centerfolds in *Playboy*. Most science-fiction writers today know this and avoid depicting aliens that are too humanoid, even though Hollywood often persists in Burroughs' tradition.

Left: An alien creature: "Contact."
Center: Six-limbed alien skeleton.
Right: "Erret skull."

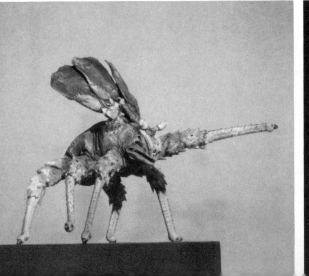

One does not have to go all the way to Mars to find strange lifeforms, however, if a little-known story of Sir Arthur Conan Doyle is to be believed.

Doyle was a physician who understood much about science, although in later life he dabbled in the occult. While his greatest creation was Sherlock Holmes, he did turn his scientific mind to science fiction. In 1913, a year after John Carter of Mars appeared, and just ten years after the Wright Brothers demonstrated heavier-than-air flight, he published a story about creatures as strange as if they were from another world, but living in our atmosphere. The story was "The Horror of the Heights" and described an "air-jungle" above southwestern England.

Although to our modern minds this idea sounds merely quaint— we know that no such creatures live up there, or if they ever did, they've all been wiped out by Boeing 747's, Concordes and Space Shuttles—Doyle's creatures eerily anticipated speculations by later scientists. The possibility has been suggested seriously by SETI scientists that planets like Jupiter could have creatures floating high in their atmospheres, suspended by bags of the lightest gas known, hydrogen.

The story takes place around 1923 (ten years in the writer's future), when the hero flies his open-air-cockpit plane with its massive 300-horsepower engine into the "upper atmosphere." There, he encounters the floating monster:

> Conceive a jelly-fish such as sails in our summer seas, bell-shaped and of enormous size—far larger, I should judge, than the dome of St. Paul's. It was of a light pink colour veined with a delicate green, but the whole huge fabric so tenuous that it was but a fairy outline against the dark blue sky. It pulsated with a delicate and regular rhythm. From it there depended two long, drooping, green tentacles,

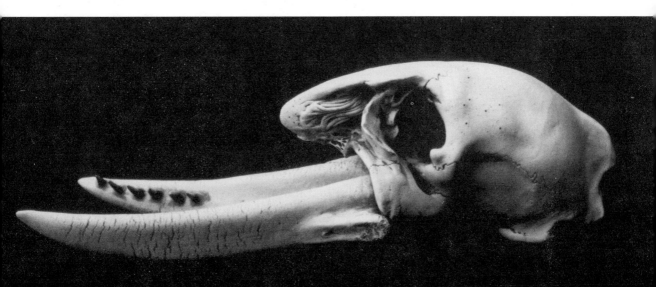

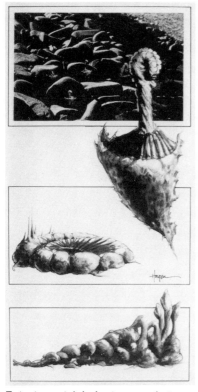

Extraterrestrial plants: xerophytes adapted to hot, arid conditions.

which swayed slowly backwards and forwards. This gorgeous vision passed gently with noiseless dignity over my head, as light and fragile as a soap-bubble, and drifted upon its stately way.

The tentacles nearly get our hero, but after a struggle, he escapes. However, his scientific spirit will not rest. He flies to the air-jungle once more to bring back proof of the marvels he has seen. His last words are:

> Forty-three thousand feet. I shall never see earth again. They are beneath me, three of them. God help me; it is a dreadful death to die!

In 1928, Philip Francis Nowlan's story, "Armageddon 2419 A.D.," introduced a man named Buck Rogers to the eager readers of *Amazing Stories* magazine. The next year, the comic strip *Buck Rogers* roared off the newspaper page, just in time to take people from the dismal world of the Great Depression into the vastly more appealing twenty-fifth century. It was the first relatively adult American science-fiction comic strip, and it became popular. So successful was it that it ran until 1967, two years before the first landing on the Moon. Its battles could have been taken from cowboy-and-Indian stories, with blasters replacing six-guns, and aliens standing in for tribal warriors, but it paved the way for more serious portrayals of extraterrestrial life.

Aliens Come of Age

Nineteen thirty-four was a landmark year in the evolution of science-fiction aliens. In *Wonder Stories*, a story appeared called "A Martian Odyssey," by a chemical engineer named Stanley G. Weinbaum. It introduced an alien named Tweel, who was *not* hostile, did *not* speak English, and did *not* even think like a human. More than that, he lived on a world with an alien ecology. The lifeforms were interrelated—not just a random assortment of monsters and vegetation designed for scenery and to provide threats to the hero. This was the most *alien* alien to date, one that matched the growing sophistication of scientific speculation about life on other worlds.

Tweel was a birdlike being who traveled by jumping high in the air and landing on his beak. But what really distinguished him from previous aliens was that he was intelligent, yet in a way different from humans. The hero's first encounter with Tweel probably foreshadows the real problems we will face if SETI is successful.

They manage to communicate each other's names, but are unable to teach each other even simple words. "I couldn't get the hang of his talk," says the human. "Either I missed some subtle point or we just didn't *think* alike—and I rather believe the latter view." This problem was solved, in part, by a technique we will probably use in our first communication with extraterrestrials:

> After a while I gave up the language business, and tried mathematics. I scratched two plus two equals four on the ground, and demonstrated it with pebbles. Again Tweel caught the idea, and informed me that three plus three equals six.

They go on to discuss astronomy, drawing pictures on the ground—essentially the technique used on NASA spacecraft decades later to provide messages for alien astronauts who may pick them up in the future.

"A Martian Odyssey" became a favorite with science-fiction fans, and literary aliens entered a new era. No longer was it such a relentlessly hostile universe. Writers depicted cooperation with other intelligent beings, rather than Darwinian battles for survival. The 1936 story "Liquid Life," by Ralphe Milne Farley, carried this to an optimistic extreme, ending with the memorable line "For he had kept his word of honor, even to a filterable virus."

Another important science-fiction writer of the 1930s who worked outside of the pulps was Olaf Stapledon. He was an Oxford-educated philosopher whose international upbringing gave him a truly universal view of the potential achievements and follies of humanity. His first novel, *Last and First Men* (1930), portrays two billion years of the future evolution of the human race, as we colonize the universe and reach higher and higher levels of wisdom.

In *Star Maker* (1937)—which has been compared in literary quality and purpose to Dante's *Divine Comedy*—the hero wanders around the universe, witnessing all manner of alien civilizations, as bizarre as anything in the pulps but conceived with a greater thoughtfulness than usually encountered in those pages. He contemplates their histories, their tragedies and their successes.

Sirius (1944) is a classic story of humans seen through an alien viewpoint, although the "alien" in this case is a dog. But he's a supersmart animal who is able to give us the nonhuman perspective on ourselves that much of the best of science fiction offers.

The aliens Stapledon created in *Star Maker* show the breadth of imagination needed to hint at the possibilities of the real universe. His awareness of the way evolution shapes all creatures, and his cosmic viewpoint (like that of SETI), makes us see our mundane selves as the product of a very specific environment. He makes us

"Fringewing Flier."

aware that most of us share an unspoken conviction that everything is the way it is because that's the only way it could ever be, although nothing could be further from the truth.

He depicts one civilization that is built by shellfish that float like boats in their shells, with sails to propel them. They are described as "living ships":

> It was a strange experience to enter the mind of an intelligent ship, to see the foam circling under one's own nose as the vessel plunged through the waves, to taste the bitter or delicious currents streaming past one's flanks, to feel the pressure of air on the sails as one beat up against the breeze, to hear beneath the water-line the rush and murmur of distant shoals of fishes, and indeed actually to *hear* the sea-bottom's configuration by means of the echoes that it cast up to the under-water ears. . . .

Or consider an even more alien Stapledon lifeform, birds that communicate through radio waves. Individually they are incapable of anything extraordinary, but together, bird-clouds become a conscious superbeing:

> . . . each one of these mobile minded clouds of little birds was in fact an individual approximately of our own spiritual order, indeed a very human thing, torn between the beast and the angel, capable of ecstasies of love and hate toward other such bird-clouds, capable of wisdom and folly, and the whole gamut of human passions from swinishness to ecstatic contemplation.
>
> Probing as best we could beyond the formal similarity of spirit which gave us access to the bird-clouds, we discovered painfully how to see with a million eyes at once, how to feel the texture of the atmosphere with a million wings. . . .

Stapledon has come, I suspect, the closest of any human to glimpsing the possibilities of the universe.

Modern Aliens

In 1945, as the bloodiest war in history was ending, a tale was published that became the classic story of the first meeting between humans and aliens. Its very title, "First Contact," has become the catch phrase to describe the meeting with other beings which we hope will one day take place in the real universe.

The story was by Murray Leinster, and described the meeting of two spaceships, one from Earth and the other from a nonhuman civilization. They spot one another while exploring the Crab Nebula, a real astronomical object thousands of light-years away.

Each ship's crew is suspicious of the other; each wants to return home, but is afraid the other will follow it home and return with invaders. It's a Mexican standoff in space. They establish communication by radio, and build a vocabulary of mutually understandable words. They find similarities in each other's thinking, but the only solution to the dilemma seems to be to fight to the death.

Then a crewman has an idea:

> "Swap ships! . . . We can fix our instruments so they'll do no trailing, he can do the same thing with his. We'll each remove our star maps and records. We'll each dismantle our weapons. The air will serve, and we'll take their ship and they'll take ours, and neither one can harm or trail the other, and each will carry home more information than can be taken otherwise!"

After the aliens agree to this, a crewman meets with one of the aliens and finds them not so different after all. They spend their last hours together swapping dirty jokes.

In those postwar years, another writer turned us back to Mars, creating images that have haunted many a reader. Ray Bradbury's *Martian Chronicles* (1950) took us to a Mars where human colonists find the remnants of a dying civilization:

> And out of the hills came a strange thing.
> It was a machine like a jade-green insect, a praying mantis, delicately rushing through the cold air, indistinct, countless green diamonds winking over its body, and red jewels that glittered with multifaceted eyes. Its six legs fell upon the ancient highway with the sounds of a sparse rain which dwindled away, and from the back of the machine a Martian with melted gold for eyes looked down at Tomás as if he were looking into a well.

Unlike so many of their literary predecessors, these Martians turn out to be friendly—it's the humans who are too often the bad guys.

Electronic Aliens

In 1934, *Flash Gordon* and his arch-enemy, Ming the Merciless of Planet Mongo landed on the pages of the Sunday comics and gave Buck Rogers some competition. Both Buck and Flash then took to the radio waves, and by 1938, the country was ready for the ultimate in radio realism: Orson Welles' broadcast of *War of the Worlds*.

Through the 1920s and '30s, television was experimented with. World War II temporarily killed its commercial development, though the great wartime advances in radar electronics made the time ripe for TV to take off when peace returned.

In 1946, the year after the war ended, and the year H. G. Wells died, NBC began the first regular television network transmissions to Philadelphia and Schenectady on the now nonexistent channel 1. Three years later, *Captain Video and his Video Rangers* shot onto the tube to defend the world against the same sort of interplanetary villains that Flash and Buck had battled on the radio. And as the Captain fought such enemies of mankind as Mook the Moon Man and Kul of Eos, *Space Patrol* and *Tom Corbett—Space Cadet* arrived to fight him for the airwaves. Despite his $25-a-week special-effects budget, the Captain outlasted his competitors, aborting his mission only when the DuMont network crashed in 1955.

The early 1950s saw a flurry of science-fiction films. Hollywood is usually decades behind written science-fiction in its sophistication, but movies began to move in the right direction. In 1950, the motion picture *Destination: Moon* gave a more realistic portrayal of traveling to the Moon than had ever been filmed. It was from a novel by science fiction master Robert Heinlein, who co-authored the film's script, and whose book *Space Cadet* was the source of *Tom Corbett*. Made with the technical advice of German rocket

"Xerophyte" plant life near a Jupiter-type planet.

pioneer Hermann Oberth, it was probably as true to science as film could be at the time. There were no Moon maidens or people breathing in a vacuum, and it did a creditable job of anticipating what would eventually become *Apollo* history.

The times were turbulent and the films often reflected this. The year after *Destination: Moon*, there appeared a memorable film that was a milestone in the treatment of aliens: *The Day the Earth Stood Still*, with which I opened this chapter. With the Korean War raging so soon after World War II, and the spread of nuclear weapons a reality, it is not surprising that someone had the idea of aliens invading the Earth to tell us to mend our violent ways (or incur the wrath of the cosmic police). Similar hopes are often expressed among advocates of SETI.

What makes this film stand out, though, is the intelligently written script by Edmund H. North (from a story by Harry Bates), and the sensitive handling by the director, Robert Wise, who also directed *West Side Story*. Their alien, Michael Rennie, comes here on a trip of friendship, as an emissary from a galactic civilization. What a change from *Buck Rogers* and *Flash Gordon*!

Sam Jaffe, as the wonderful old professor in this film, conveys the essence and spirit of the true scientist when he realizes he is in the presence of the extraterrestrial. The professor's eyes light up excitedly and he realizes that the answers to all the questions of science can be had by conversing with this visitor from a vastly more advanced civilization. This is the spirit that drives most SETI scientists today, even if some might not admit it publicly.

The flip side of this story was seen in many other films. If we had the word of a United States Senator—Joseph McCarthy—that our very own government was filled with Communist spies and saboteurs, then what could be more natural for Hollywood than to create sneaky aliens who are able to disguise themselves as your friends until you're alone with them and at their mercy?

This was the theme of such movies as:

• *The Thing* (1951). Produced by Howard Hawks, from the story "Who Goes There?" by the legendary writer-editor John W. Campbell, Jr. Scientists dig up an evil, frozen alien (the pre-*Gunsmoke* James Arness) who thaws and terrorizes an Arctic base. In the original story, it was capable of taking the form of any human. Which one of your friends is *it?* Or is it *you?* (Remade in 1982.)

• *Invaders From Mars* (1953). This one is about hulking green Martians who program everyone's mind to pave the way for a complete takeover. It gave me nightmares when I was a boy, even though (or perhaps because) I lived then at White Sands Missile Range, New

Mexico, where my dad routinely launched V-2 missiles into the edge of space while most people blissfully dismissed space travel as Buck Rogers nonsense. (Remade in 1986.)

• *Invasion of the Body Snatchers* (1955). Directed by Don Siegel (who also directed *Dirty Harry*) from a book by Jack Finney, with some uncredited contributions by Sam Peckinpah. Giant seed-pods from space land, and duplicates of townspeople start to grow inside them, taking the place of their alter egos when they sprout. Before long, Kevin McCarthy, the only human left in town, tries to escape to warn the world. So worried was the studio about the effect of an ending which shows McCarthy unable to alert the world that they forced the director to tack on a different ending saying, in effect, that the word got out and humanity was saved. Mercifully, the second ending was cut from later releases.

Paranoia was not the only theme explored in films of that era. Two fine pictures explored more positive attitudes toward alien civilizations: *This Island Earth* (1954) and *Forbidden Planet* (1956). In both, humans travel to planets of other stars and encounter basically good civilizations that are far beyond ours technologically, though not psychologically: they have not found a cure for violence.

Life Imitates Art

In 1957, reality overtook science fiction: the Russians launched *Sputnik* and the Space Age arrived. Just two years later, two scientists, Giuseppe Cocconi and Philip Morrison, proposed searching for radio signals from other civilizations in the galaxy. Independently of their suggestion, such a search was actually begun the following year by radioastronomer Frank Drake.

In 1959, the television industry caught up with the Space Age. It proved very difficult to convince the network and advertisers to back a science-fiction series, but luck and talent prevailed and soon, each week, a television host stepped into our living rooms with these words:

> There is a fifth dimension beyond that which is known to man. It is a dimension as vast as space and as timeless as infinity. It is the middle ground between light and shadow, between science and superstition, and it lies between the pit of man's fears and the summit of his knowledge. This is the dimension of imagination. It is an area we call the Twilight Zone.

Thus writer Rod Serling introduced *The Twilight Zone*, whose very name entered the language as a synonym for the possibilities of the universe.

Two years later, a Russian, Yuri Gagarin, flew into the vastness of space and became the first human to orbit the Earth. Three months after that, science stepped into the Twilight Zone of politics as President John Kennedy made one of the most extraordinary decisions in history and announced:

> ... I believe that this Nation should commit itself to achieving the goal, before this decade is out, of landing a man on the moon, and returning him safely to earth. No single space project in this period will be more exciting, or more impressive to mankind, or more important for the long-range exploration of space; and none will be more difficult or expensive to accomplish.

At about the same time, inspired by the success of *The Twilight Zone*, television presented *The Outer Limits*, whose host intoned each week:

> There is nothing wrong with your television set. . . . We are controlling transmission. We will control the horizontal. We will control the vertical. For the next hour, sit quietly and we will control all that you see and hear. You are about to participate in a great adventure. You are about to experience the awe and mystery which reaches from the inner mind to—*The Outer Limits*.

The first episode, "The Galaxy Being," brought to Earth an alien who traveled electromagnetically to be picked up by a radiotelescope. Astronomers were now in fact using such devices to search for radio signals from other civilizations.

And then came 1966. We still hadn't landed on the Moon, but a media event occurred that will live forever in the minds of would-be astronauts everywhere. It began with these words:

> Space: the final frontier. These are the voyages of the starship *Enterprise*. Her five-year mission: to explore strange new worlds, to seek out new life and new civilizations, to boldly go where no man has gone before.

This, of course, was the first episode of *Star Trek*. It introduced to this planet the very human Captain Kirk and his ever-logical, half-human, half-Vulcan science officer, Mr. Spock. In the real universe the previous year, NASA's *Mariner 4* had become the first spacecraft to fly by the planet Mars successfully, and the Soviet *Venera 3*, the first to hit another planet (Venus). The public was certainly ready for a realistic look at interstellar travel.

Star Trek's Mr. Spock (Leonard
Nimoy), the super-intelligent alien
of the planet Vulcan, and Captain
Kirk (William Shatner) of the starship *Enterprise*.

The series was created by former pilot Gene Roddenberry, whose
experiences at the controls of airplanes strongly influenced the
design of the starship. Initially the show was *not* a big success, and
was canceled after only two seasons. But a huge letter-writing campaign by fans convinced the network to bring it back for one more
season. Its optimistic exploration of the universe has struck a chord
with so many people that it continues to be one of the most popular
series in syndicated rerun, and has inspired three motion pictures.

During the *Star Trek* era, director Stanley Kubrick was working
with writer Arthur C. Clarke to make a film based on Clarke's short
story, "The Sentinel," about humanity's first contact with aliens.
Kubrick planned it meticulously, consulting with such scientists as
Carl Sagan. The title was *2001: A Space Odyssey.*

The making of that movie was virtually a race to keep ahead of
NASA's attempt to get to the Moon. It was finally completed in 1968,
the year astronauts Borman, Lovell and Anders became the first
humans to fulfill Jules Verne's dream and orbit the Moon.

2001: A Space Odyssey was not a hit at first. It was too puzzling
and radically different from the films of the genre that had preceded
it. But word-of-mouth about its authenticity and powerful emotional
impact spread among scientists, engineers and science-fiction fans,
until the public at large caught on and gave it the success it deserved.

Sagan had convinced Kubrick not to use the usual actor-in-an-alien suit on the grounds that our own forms were such an accident of numerous random evolutionary events that the odds of aliens looking humanoid were small. This advice, along with the difficulty of making truly nonhumanoid aliens who didn't look fake, led Kubrick to keep the creatures off-screen.

It was by far the best exploration yet of the possibilities of alien contact. After a grueling trip to Jupiter, the sole surviving human astronaut travels through an alien space-warp and is given a tour of the universe. In the end, he is educated by the advanced civilization and returns to Earth reborn.

That many people were mystified by the climax is itself a testimony to the film's achievement, for the first contact with a culture that has a completely different history, and biology, raised on a different world around another sun, must be a far more jarring encounter than that of the Native Americans with the first European explorers. Much would be totally incomprehensible at first, so the disturbing inability of most people to make sense out of the climax probably is the best inkling of what we will experience if SETI succeeds.

The next year, on July 20, 1969, Neil Armstrong and Buzz Aldrin stepped onto the Moon while Michael Collins orbited around it. Visits to another world were no longer fiction.

A film race of sorts has paralleled the space race. The Russians have long enjoyed science fiction, and made one of the earliest, *Aelita*, in 1924, from a play by Alexei Tolstoy (a relative of *the* Tolstoy). What may be their best film in the field was made in 1972: *Solaris*, from a novel by Polish writer Stanislaw Lem. Interstellar astronauts visit a planet whose ocean is a gigantic alien. It's obvious that the filmmakers had seen *2001: A Space Odyssey*, and great care was taken to emulate the standards of special effects used there. Some critics consider it to be one of the finest science-fiction films ever.

For a while, reality outpaced science-fiction filmmakers: astronauts visited the Moon for the last time in 1972; NASA's *Pioneer 10* spacecraft became the first spacecraft to fly by Jupiter (1973); scientists transmitted a message to a distant star cluster (1974); Russian and American astronauts met in space and Russia put probes on Venus (1975); and *Viking 1* answered Percival Lowell's dreams by becoming the first spacecraft to land on Mars successfully (1976).

Then, in 1977, science fiction retaliated for these incursions on its territory with two films: *Close Encounters of the Third Kind* and *Star Wars*.

Close Encounters of the Third Kind was Steven Spielberg's vision of "First Contact," with aliens visiting Earth in a huge starship.

Star Wars, made by his friend and frequent coworker, George Lucas, affectionately recycled all those old space adventures from the pulps and Saturday-morning serials and served them up as a coherent, believable galactic civilization that looked far more "lived in" than any previously portrayed on the screen. For many fans, including this author, the cantina scene is one of the classics of film, wherein Luke Skywalker (Mark Hamill) and Han Solo (Harrison Ford) meet some of the sleaziest nonhumanoid aliens ever to growl from a movie screen. At last, Hollywood had freed itself from humanoid chauvinism. No longer did aliens look and sound like actors in rubber suits.

Close Encounters and *Star Wars* quickly became two of the most popular films in history. The next year, NASA's *Pioneer Venus 1* and *2* probes orbited and landed on Venus; in 1979, their two *Voyager* spacecraft flew past Jupiter and took superb pictures of the planet; the same year, one of the *Pioneers* that had flown by Jupiter also journeyed past Saturn; in 1981, the first Space Shuttle flight ushered in the age of reusable spaceships.

Then, in 1982, while NASA battled a reluctant Congress for funds to search for extraterrestrial intelligence, there appeared a film that captured in many ways the spirit of SETI: Spielberg's *E.T.* The little alien won the hearts of the world, making up for those decades of nasty film monsters, and incidentally entering the *Guinness Book*

Steven Spielberg's E.T.

of World Records as the most popular movie in history. The popularity of the film may have helped some Congressmen to perceive the political value of supporting SETI research.

In 1985, Carl Sagan stepped across the boundary between science and fiction with his first novel, *Contact*. The book depicted a vision of the first contact with extraterrestrials as a consequence of the very SETI that he had helped to establish.

Fiction into Science

If the influence of science on fiction has become clear, let us not forget that the reverse path has also been followed. Wernher von Braun was inspired by the German film *Die Frau im Mond* (released in English as *The Girl in the Moon*) to make lunar rockets a reality. Carl Sagan traces some of his interest in the exploration of the cosmos to John Carter of Mars. And they are not exceptions; many scientists, engineers and astronauts have been inspired to make science fiction into fact. We are now living in the world they created, and the possibility that fictional aliens may join the long list of literary inventions-come-true is the subject of the rest of this book.

Even if we now know there are no Martians, the same technology that allowed us to send robots to Mars in search of them will eventually enable us to send people. One day, I am convinced, there will be humans living on Mars. They may even build covered canals to make the nineteenth-century Mars a twenty-first-century reality. One day we may see Ray Bradbury's *Martian Chronicles* description of such colonists come true:

> "I've always wanted to see a Martian," said Michael. "Where are they, Dad? You promised."
>
> "There they are," said Dad, and he shifted Michael on his shoulder and pointed straight down.
>
> The Martians were there. Timothy began to shiver.
>
> The Martians were there—in the canal—reflected in the water. Timothy and Michael and Robert and Mom and Dad.
>
> The Martians stared back up at them for a long, long, silent time from the rippling water. . . .

In the Beginning

THE ORIGIN OF LIFE

Science is a flickering light in our darkness, it is but the only one we have and woe to him who would put it out.
Morris Cohen

IF THERE WERE a *Cosmic Cookbook of Life*, written by some galactic Julia Child, it would probably read something like this: take one medium-size planet, place it near an average star. Add water, ammonia, methane, perhaps a dash of carbon dioxide. Turn on the electricity in the form of lightning. Stir. Wait patiently—a few hundred million years will do. Then something will start swimming around in your soup. You may enjoy eating it. Or vice versa.

That is a rough summary of the knowledge we have gained about how we think life arose on Earth. Of course, it doesn't mean that's how it must arise elsewhere, but until we understand how life arose here, it's impossible to estimate the odds of its developing on other worlds.

So in order to consider SETI intelligently, let's look at what is known about life's origin, and what astronomy tells us about the possibility that the ingredients in our recipe for life might be available in the galaxy.

Genesis, Revised

When we look in the bathroom mirror, we see biology. To understand how biology could arise from lifeless substances forces us to look at geology. To understand where the rocks came from, we have to turn to astronomy. And that leads to the question, where did the stars come from? And so the quest to understand our ultimate roots leads us step by step further back in time and farther away in space until it encompasses the history and geography of the entire universe.

An extraordinary amount has been learned in this century about the origin of life on Earth. Chemists, biologists and biochemists have studied the complex chemicals of which living things are made, and have learned how to build complex biological molecules from simple ones.

At a 1985 Caltech conference on the origin of life held in honor of Norman Horowitz, one of the major contributors to the field, Horowitz said:

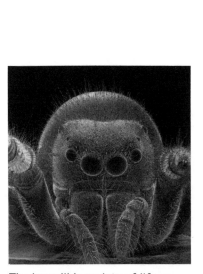

The incredible variety of life on Earth is hinted at by this scanning-electron-microscope picture of an eight-eyed jumping spider.

> People sometimes ask me, "What can you expect to learn by studying the origin of life? After all, we can't know what *really* went on." I agree that this may be true, but I think it's important, nevertheless, to have at least one plausible scenario, one plausible route, to the origin of life—one that meets all the constraints imposed by chemistry, geology, and astronomy.
>
> I don't think one can claim to have a really deep understanding of life if one doesn't know how it might have begun. I read a recent article by Steven Weinberg in . . . *Science,* in which he said that cosmologists are now hoping to discover the origins of the physical laws that govern the universe. Well, compared to that goal, I would say the search for the origin of life is rather modest.

The assembled scientists laughed. He continued, "Of course, it may turn out that no realistic scenario is possible once the geologists and planetologists finish working over the history of our planet. It may turn out that conditions were unsuitable for synthetic organic chemistry." Horowitz points out that if we were ever forced to conclude that life could not have arisen on Earth by natural processes in the time available, we would then have to start looking for its

A tiny sample of the diversity of life on Earth, each of which has evolved to fill a different environmental niche.

origin on other worlds and imagining ways that it could have gotten here.

Geologists have read the records written in the rocks, deciphering fossils and minerals to tell the history of life's evolution. Not only do fossils contain skeletons of ancient creatures and plants, but they even contain clocks—natural clocks in the form of radioactive atoms.

Almost everything in the universe is slightly radioactive. This means that in any rock you pick off the ground, the nuclei of some atoms are breaking down every second, spitting out electrons, protons, gamma rays or other particles. By comparing the amounts of unaltered atoms and changed ones, scientists can tell how old the rock is. They can calculate the date when the rock was born, in other words, when it cooled from its original molten state.

Archbishop Ussher, by taking the Book of Genesis literally, calculated that the world was born in 4004 B.C. Geologists find, however, that the Earth is almost five billion years old, and they get similar figures from studying meteorites and lunar samples.

Physicists and geologists have created theories of how stars and planets form. Astronomers have found many stars similar to our Sun, and have even observed stars that seem to be forming now. They have found evidence that the whole universe was born in a stupendous explosion, commonly called the Big Bang.

Why we know the universe is a lot older than 6,000 years: uranium decays radioactively by emitting particles. After emitting several particles, it becomes the metal lead (chemical symbol *Pb*, from *plumbum*, Latin for lead). For the most common form of uranium in nature, the half-life for this process is about 4 billion years, meaning that after 4 billion years, half the uranium atoms become lead; after another 4 billion years, half of the remaining uranium becomes lead; and so on, forever. The example shows what happens when you start out with four atoms of uranium. After 4 billion years, two of them have become lead. After another 4 billion years, one more atom becomes lead. Thus if you find a rock that is half uranium and half lead, and it has not been chemically altered by nature, you know it is 4 billion years old, which is about the age of the oldest known rocks on Earth.

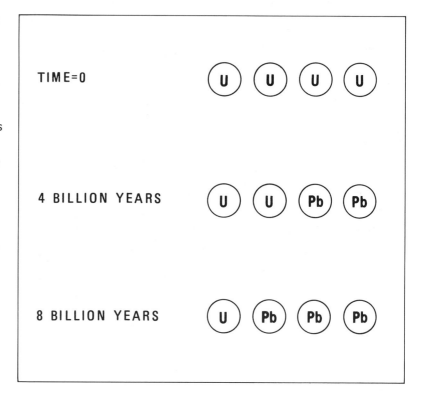

By putting all this information together, we have a remarkably consistent picture of how our own world was born, and how life arose. That there are many gaps in the picture and conflicting theories of particular chapters in the book of life should not be as surprising as the amount of knowledge we *do* have. Earthquakes, volcanoes and meteorites have violently obliterated some of the records; wind, water and the drift of continents have slowly erased many more. Yet what remains provides numerous clues, and new information about the Moon, asteroids, comets and other planets is filling in some of the gaps.

The Book of Genesis, if written by scientists, would read this way:

In the beginning a Big Bang created the heaven and the Earth.

And the Earth was without form and void; and the tiniest of elementary particles were upon the face of the universe.

And the Big Bang exploded and there was light, and X-rays and all manner of photons, and it was good.

And as the universe expanded, tiny particles formed bigger particles, until electrons cleaved unto protons and begat hydrogen atoms.

And the electrons and protons and hydrogen gas which was the universe expanded, and as it expanded, cooled. And as it cooled, gravity manifested itself everywhere and drew together clouds of gas which formed giant swirls called galaxies.

And within each galaxy, gravity drew gas and dust into clumps, and these clumps were called stars, and there were lights in the firmanent of heaven.

And around those stars, gravity drew in smaller clumps and these were called planets; and yet smaller clumps were called moons, asteroids and comets.

And after billions of years had passed, a star was born which was called the Sun. And around this star turned nine planets, the third of which was called the Earth.

And as the Earth cooled, water rained from the heavens, and oceans divided the land by the gathering together of the waters called the seas.

And the air was ammonia and methane and hydrogen and other gases, and lightning did thunder mightily. And the chemicals did mix within the seas, and the lightning and the sunlight did create more complex chemicals until was born a molecule which duplicated itself.

And this molecule begat more molecules. And each molecular son begat yet more molecules, generation upon generation. The laws of chemistry said: Be fruitful and multiply.

Sometimes, the new molecule was slightly different from the old. Thus it came to pass that a new molecule was born that was better than its parent—better able to use its environment to survive.

And thus were born ever more complex molecules, until groups of molecules came together which were called cells, and it was called life.

And groups of cells gathered together to form yet more complex creatures, and some were called the fish of the sea.

And chemistry created great whales, and every living creature that moveth, which the waters brought forth abundantly, after their kind, and every winged fowl after his kind.

And plants did grow upon the Earth, and creatures of the sea did conquer the shores. And their descendants did flourish upon the land. And it was good.

And that, dear reader, is how we think you came to be here.

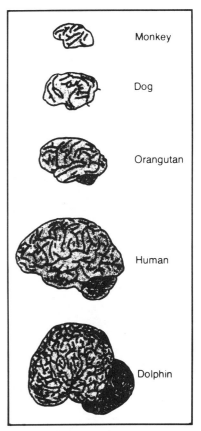

A comparison of brain sizes. The dolphin brain is larger than ours, though it does not seem to be as smart. Brain size is a function of both intelligence and body size. Large bodies need large brains, so the elephant brain is several times larger than our own, but not nearly as clever. (From *Cosmic Search*.)

HERE'S THE PICTURE, BOYS—FEATHERS, WINGS, BEAKS, AND NESTS. IT'S A CRAZY IDEA, BUT IT JUST MIGHT WORK!

Cosmic Roots

If Alex Haley, the author of *Roots*, had traced his ancestry back far enough, he would have had to make room in his genealogy chart for a star. Many of the atoms in his body were once inside a sun. How do we know this? It is from a chain of deductive reasoning worthy of Sherlock Holmes. One of the clues is encoded in the light from other galaxies.

We live inside of one such galaxy, the Milky Way, with hundreds of billions of stars orbiting around the center. The whole Milky Way galaxy is about 100,000 light-years in diameter. (A light-year, being the distance light travels in a year, equals six trillion miles.)

For comparison, our Sun's nearest neighbor, the Alpha Centauri star system, is around four light-years away. We live about 30,000 light-years from the center of the Milky Way, so we are actually closer to the rim of our galaxy than to the center. We're out in the galactic suburbs. At this distance, it takes us a quarter billion years to orbit around the galaxy once.

At night, if you get away from the city lights and smog and look up, you see a smooth white band across the sky that the Romans called the *Via Lactea*, the Milky Way. It consists of billions of stars, and it is just our galaxy viewed from the inside looking out. The Greeks called it *galaxias kuklos*, the milky circle, giving us the word for galaxy.

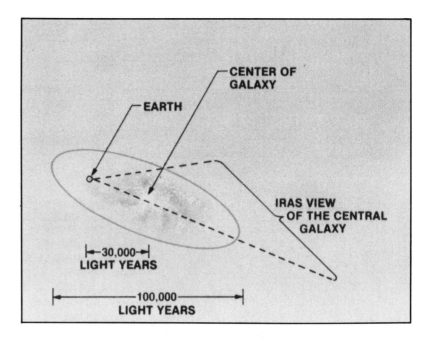

You are here. Earth is located about 30,000 light-years out from the center of the disk-shaped Milky Way galaxy. (IRAS is the Infrared Space Telescope.)

In a telescope, many fuzzy dots in the sky are revealed as similar swirls of stars, galaxies. In fact, there are billions of galaxies, each containing billions of stars. This is the single most important reason for optimism in the search for life. In a universe so vast, with stars as numerous as grains of sand, it is hard to imagine that the conditions for life have not arisen elsewhere.

Because of the enormous distances, it might seem impossible to learn any details about galaxies, but fortunately, their light contains information written by the stars. If you pass sunlight through a prism, you get a rainbow of colors, called a spectrum (rainbows are produced this way, with droplets of water in the air acting as tiny prisms).

The spectrum of a star has bright and dark bands caused by atoms in the stars. Iron atoms, for example, absorb certain narrow bands of colors, producing a distinctive pattern called the iron spectrum. Hydrogen, silicon, calcium and, in fact, every chemical element and compound has a unique "fingerprint" of color bands, a distinctive spectrum that allows it to be identified in sunlight. Thus we can tell what stars are made of, and we find that they are made of the same elements we find on Earth.

"There goes the ecology."

People sometimes wonder how we can presume to predict the origin of life when the laws of the universe could be different elsewhere. The details of the spectrum of starlight—how intense one line is compared to another, what the precise wavelength (color) is, and so forth—are critically dependent on the laws of physics. Everywhere we look in the universe, we find the same laws operating. Perhaps one day in some mysterious black hole, or at the farthest edge of the universe revealed by the Hubble Space Telescope, we may find some clue that forces us to revise our laws. But until then, we have mountains of evidence that the same basic rules of physics that govern life here operate everywhere else. So we have some confidence that the lessons we learn about life on Earth may apply to other stars.

On a starry night, photons—particles of light—constantly rain down onto your body. Many of them are from stars hundreds of light-years away, so the photons you see are centuries old. In the constellation of Andromeda is a nearby galaxy bright enough to be seen with the naked eye as a fuzzy patch of light. The Andromeda galaxy is so far away—two million light-years—that the light hitting you left when your ancestors were just beginning to evolve into humans. The sky is a time machine allowing us to peer into history. The next time you see the stars, think: you are being touched by light which left when Julius Caesar ruled; when the Pyramids were built; when

GALAXY NGC-7619 IS MOVING AWAY FROM US AT 3800 KILOMETERS PER SECOND, AND YOU WANT THE AFTERNOON OFF!

"I love hearing that lonesome wail of the train whistle as the magnitude of the frequency of the wave changes due to the Doppler effect."

the wheel was invented; when your ancestors were just beginning to evolve into humans.

And up there, some creature may be looking at the sky toward you.

Each of those photons is a tiny history book, a book that can be read with the aid of telescopes and spectroscopes and thought. Not only does the light tell us what chemicals make up each star, and what its temperature is (because the intensities of lines change with temperature), but that light even tells us the speed of each star as it travels around the galaxy.

The colors of a star's spectrum change slightly if the star moves. They become redder if it moves away from us, bluer if it moves toward us. This is the phenomenon of the Doppler effect, the same one that causes an automobile horn to sound higher pitched when it is moving toward you and lower pitched when moving away. Galaxies, we find, are red-shifted. The galaxies are moving away from us.

Why? Most scientists think the explanation is that there was a big explosion, the Big Bang, when the universe was born about fifteen billion years ago. Thus galaxies are moving away from one another, like fragments of a hand grenade.

Powerful evidence for this Big Bang came from a most unlikely source: Ma Bell. This is a beautiful example of the way in which mundane activities can lead to scientific breakthroughs.

In 1963, six years after *Sputnik*, two scientists, Arno Penzias and Robert Wilson of the Bell Telephone Company, were studying radio interference of communication satellites. They found that the noise was coming from all over the sky. It was a faint hiss, at frequencies similar to those used in microwave ovens.

Penzias and Wilson had no idea what caused it, but they proved it was coming from space. They asked Princeton physicist Robert Dicke about it, and it just so happened that his research team was building a device to search for the microwave noise that ought to have been produced by the Big Bang, if it had really occurred. He immediately realized that Penzias and Wilson had found that cosmic noise.

As a result, Penzias and Wilson received the Nobel prize. Had they been less diligent in their research and just written the noise off as some minor interference, Dicke would probably have discovered it and received the Nobel prize instead. Such are the unpredictable byways of research.

To most scientists, this cosmic background noise is proof of the Big Bang theory. Just as fossils tell us about the history of life on Earth, this noise is a radio fossil of the dramatic history of the universe.

But what does that tell us about how we are made? Well, in the beginning, the universe was one huge blob of exploding matter. As it expanded, it cooled, and atoms formed. Physicists calculate that most of the atoms which formed would be the simplest of all, hydrogen: a single electron orbiting around a proton. How could the more complex atoms such as carbon or iron arise?

They had to be "cooked" up. The first stars were mostly hydrogen. But stars burn by fusing atoms together, building more complex ones from simple ones, the same process that gives the hydrogen bomb its power. After a star burns for millions or billions of years, it converts much of its original hydrogen into more complex, heavier atoms. They would be stuck inside the star forever if stars didn't

A star is born. This is an infrared photograph of a star forming, as seen by the IRAS space telescope. Gas and dust are pulled into a clump by gravity, forming the hot, bright spot. It may be millions of years before the star has fully formed.

sometimes explode. In any given galaxy, however, several times a century, a star explodes as an incredibly violent supernova. For a few days, the supernova may be as bright as all of the hundreds of billions of stars of a galaxy put together!

The supernova spills its guts into the galaxy, shooting the cooked atoms out into space where they can be picked up by stars that are just forming. The shock wave triggers the birth of stars as the gas and dust of space are compressed. The formation of the solar system itself may have been triggered by a supernova.

Our own Sun and planets were produced after the galaxy had been operating for billions of years. We were formed from the debris of the cosmos, much of which had been cooked inside other stars. We are made of recycled stars.

Look at your hand. Ponder for a moment that every bit of flesh you see was once inside the most stupendous explosion the universe has ever known, the Big Bang. And much of your body was once inside a supernova. In a very real sense, you were born inside a star.

Interstellar Goodies

One of the most important discoveries in the search for extraterrestrial life came about through the same type of device that Pen-

A photograph in visible light of one of the nearest galaxies, Andromeda, about two million light-years away. If we could see our own Milky Way from the outside, this is roughly what it would look like, with hundreds of billions of stars.

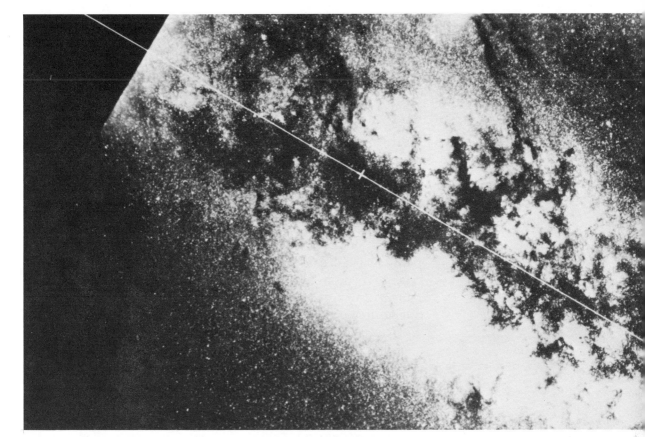

The center of our galaxy, crowded with stars, here seen in visible light. The diagonal line is inserted to show the plane of our galaxy; the dark splotches are the dust that keeps us from seeing the very center. (The black corner is just the edge of the photograph.)

zias and Wilson used to discover the cosmic background noise: the radiotelescope. Typically, a radiotelescope is a dish antenna similar to those used to pick up television signals from communication satellites, operating at similar frequencies.

If the space between stars were a perfect vacuum, it would be the end of this part of the story. But interstellar space is actually filled with a very thin gas, thinner than the best vacuum we have made on Earth. Typically, there are around ten atoms per cubic inch between the stars. But space is so vast that even this near-vacuum can add up to enough matter to make a star.

We have known for decades that there were a few types of atoms and simple molecules in interstellar space, because the spectrum of a star shining through an interstellar cloud shows atomic and molecular lines. But we thought that more complex molecules couldn't exist in space because of the harsh environment of ultraviolet starlight, which breaks molecules apart. Then radio astronomers began studying microwave noise from space and found that there are huge clouds, light-years in size, of astonishingly complex molecules—including some of the building blocks of life.

Just as atoms produce distinctive patterns of absorption and emission of light, molecules absorb and emit microwave radio energy in similar patterns. In the last decade, dozens of molecules have been found in interstellar space, and the list keeps growing. As techniques become more sophisticated, we find bigger and bigger molecules, and there seems to be no limit to their size. Apparently, the reason they survive the rigors of space is that dust clouds shield them from harmful ultraviolet, and dust grains roughly the size of bacteria provide solid surfaces where atoms and simple molecules can stick to others, building up more complex molecules. In effect, the dust grains are molecular factories, or microscopic planets.

Water (H_2O), hydrogen (H_2), ammonia (NH_3) and methane (CH_4), or their related compounds—the primary ingredients in the recipe for life—have been found in interstellar space. A total of about sixty different molecules has been identified so far, and the main limiting factor seems to be on Earth, due to the difficulty of measuring the microwave spectra of heavy molecules in the laboratory. Until such measurements are made, we cannot identify some of the molecules detected by astronomers. Understanding this great pharmacy in the sky has led to the development of a whole new branch of chemistry called *galactochemistry*.

Most of the molecules found in space are technically "organic." This does not mean that they are necessarily connected with life, but simply that they are carbon compounds. Carbon is the basis of all the compounds of life. Hydrogen, nitrogen and other atoms stick to carbon, carbon atoms stick to each other, and chains of enormous complexity form. For example, the blueprints of our bodies are contained in the DNA molecules within each cell, and DNA is basically just a gigantic carbon molecule.

During our Sun's wanderings around the Milky Way, we must have occasionally passed through interstellar molecular clouds. Perhaps some of the chemicals necessary for life were provided this way. If so, life may have been similarly seeded on other worlds.

Regardless of the details, we now know for sure that the basic chemicals of life are widely distributed throughout our galaxy. This is one of the most important reasons for optimism in SETI: with the basic ingredients so widespread, how can life be rare?

Son of Frankenstein

Can we create life in the laboratory if we start with these basic ingredients? Not yet, but we can make some major steps in that direction.

The way was first pointed by the Russian scientist Aleksandr Oparin. Oparin graduated in 1917 from Moscow State University, where he studied the physiology of plants under an acquaintance of Charles Darwin. It was the year of the Russian Revolution, and Oparin was soon to create his own revolution. In 1922, at a meeting of the Russian Botanical Society, he presented his theory of the origin of life.

Oparin proposed that life first arose in the oceans, in the chemical soup that would have existed on the early Earth, and that sunlight energized the reactions. He published his ideas in Russian in 1924, but it was not until they were translated in the 1930s that they began to influence the scientific world at large, when they merged with the thinking of British biologist J. B. S. Haldane.

It was not until 1955 that the first major experiment testing the Oparin-Haldane theory was performed. Stanley Miller, a graduate student at the University of Chicago, working under chemist Harold Urey, put what they hoped were the ingredients of life into a bottle: ammonia, methane, hydrogen and water. In a scene that might have come from *Frankenstein*, Miller ran an electric arc through the mixture, simulating lightning. After a week, he analyzed the results. To his delight, there were amino acids, the building blocks of proteins.

Many similar experiments have been run by scientists since then, and they generally succeed in producing a menu of organic compounds even more complex than those so far found in interstellar clouds, many of which are known to play vital roles in living things.

This gives us courage to attempt to answer the question, would life on other worlds resemble ours? Carl Sagan, at a SETI meeting, said, "I think the one thing that's *guaranteed* is that life elsewhere will not be like life here—it's just the nature of the evolutionary process. There's convergent evolution, of course, but different environment, different chemistry, different evolutionary sequences. It would be *astonishing* if anybody who looked very much like you or me arose elsewhere."

Controversies and Possibilities

I must not leave you with the feeling that we have completely unraveled the story of life. The story is still incomplete, with many puzzles to be solved, and numerous controversies enliven the research.

Just in the last few years, an extraordinary new idea has taken hold. Luis Alvarez, a Nobel prizewinning physicist, together with his son Walter and several colleagues, proposed that much of the

As alien a monster as ever populated the pulps. In reality, this is a tiny grain weevil emerging from a grain of barley.

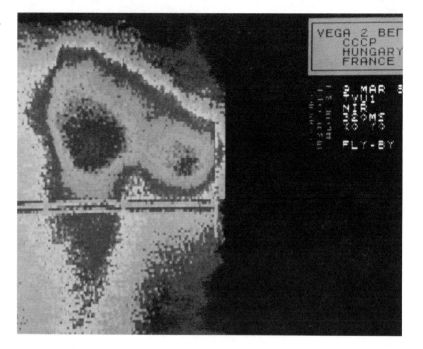

A *VEGA 2* picture of Halley's Comet, taken as the Soviet spacecraft approached. The two bright spots are jets of gas and dust spewing forth from the comet; the two dark horizontal lines are noise. Many of the cometary dust particles that hit the spacecraft were made of carbon, nitrogen, oxygen and hydrogen—the ingredients of life.

"The real reason dinosaurs became extinct."

life on Earth was wiped out 65 million years ago when an object from space collided with our planet.

They found evidence in a 65 million-year-old layer of Italian soil that an asteroid—a mountain of rock from space—or a comet hit the Earth and wiped out many species, including the dinosaurs. A collision with a mountain-sized object would have spewed staggering amounts of debris into the air, covering the Earth with thick clouds, blocking sunlight for months or years and killing many plants. Temperatures would have plummeted. Animals would have starved and frozen.

Controversy still rages over this theory. All agree that Earth has suffered severe impacts from space, but whether these caused the extinction of life is still passionately debated. I find the evidence for it compelling, as do many others, though quite a few are still unconvinced. Certainly, we now must take seriously the idea that occasionally, throughout the history of life, cosmic debris has struck our planet in collisions of more energy than in all of the bombs ever built.

This is forcing us to rethink some details of evolution. It may be that evolution proceeds slowly except when the planet undergoes a cataclysmic collision—or when a series of enormous volcanic eruptions is triggered (the major competing theory of mass extinctions of life). Then many species die, and previously insignificant creatures take over, as their worst predators are no longer around to bother them.

Even more controversial is the theory that there is an as yet undetected, faint star orbiting around our Sun. Called Nemesis, this star would take about 30 million years to orbit around the Sun. According to this theory, every 30 million years or so, Nemesis comes fairly close to the Sun (though still beyond the planets), stirring up the billions of comets orbiting just outside the planets. Some of those comets come roaring through the solar system, crashing into planets with disastrous results. Not many scientists accept this theory, but Nemesis is now being searched for.

Then there are controversies concerning the origin of life. There is, for example, the theory of A. G. Cairns-Smith, a chemist at Glasgow University, Scotland, who believes that the first crucial chemical steps took place not in the oceans but in clay. He thinks molecules evolved on the surface of crystals within the clay. If he is right, then, as he writes in his book *Genetic Takeover*, "The startling conclusion is that our first ancestors were literally made of clay."

A most unorthodox British astrophysicist, Sir Fred Hoyle, has provided an assortment of controversial ideas. Hoyle developed many of our concepts of the formation of the chemical elements as outlined earlier. He is also the author of numerous science-fiction novels. But in addition to his widely respected astrophysical research, he has produced a series of books with his colleague, Chandra Wickramasinghe, that challenge conventional thinking about life.

In *Lifecloud*, they propose that life originated in interstellar space, not on Earth, and was brought here on comets like manna from heaven. In *Diseases From Space*, they suggest that plagues and diseases also have fallen to Earth from such comets. And in *Evolution From Space*, they present the idea that life was seeded on Earth by an advanced intelligence. They argue that there were too many steps in the origin of life to have taken place by random Darwinian evolution during the mere five billion years available on Earth. Life is too complicated, they say. The odds of randomly assembling the particular sequence of atoms in our DNA molecule are very poor, they think. In this view, life began somewhere else in the universe, and was brought here in some primitive form which was then modified by conventional evolution.

Then there is the British scientist Francis Crick, and his colleague Leslie Orgel, who also think life may have been seeded from space by an extraterrestrial civilization. Crick is well aware of how unbelievable the idea seems to most scientists. "How *could* such

DNA embraces
the planets.

stuff be considered seriously?" he writes in his book *Life Itself*. "The
whole idea stinks of UFOs or the Chariot of the Gods or other common forms of contemporary silliness." Yet he thinks it could be the
true story. Crick is not easy to dismiss. He is the Nobel prizewinning codiscoverer of the DNA molecule's double-helix structure.
Crick and Orgel, too, think the chances of evolving enormously complex creatures like us from nonliving chemicals by random evolution are just too small to have been possible in five billion years.

My own opinion—and most other scientists would agree—is that
it is too early to jump to such fascinating but artificial explanations as Hoyle and Crick propose. There are many details in the
history of life that are unclear, but progress in unraveling the

mysteries is routinely made in laboratories all over the world. If there is a lesson to be learned from our experience thus far, it is how remarkably *easy* it is to make the chemical building-blocks of life. These are only the first steps toward life, and many more stages remain unknown before we can say, "start out with molecules A, B and C, follow the recipe, and out will come a molecule that splits in two, grows into duplicates of itself and splits again—the first living chemical."

But if we can do so much so easily with experiments lasting no more than months, imagine what can be done with five billion years and a planet-size vat of chemical soup.

If you shuffle a pack of cards, you get a random sequence of aces and deuces and all the other cards. The chances of getting that particular sequence of fifty-two cards were infinitesimal, but there it is. If we could go back and start evolution again, we would probably get a different sequence of chemicals, a different series of atoms in the DNA molecule, and different creatures. There might be pigs with wings and octopi swinging in the trees; lizards might rule the Earth. Until we know more of the details of life's history, we cannot say for sure whether life was a fluke or a certainty.

But there are some things we can say confidently. One of the most encouraging facts about the search for life is that we are made of some of the most abundant atoms in the universe: hydrogen, carbon, oxygen and nitrogen. If we had been made of something rare, like platinum, we might have more reason to be pessimistic about the chances for life.

Many meteorites from space have even been found to contain some of the organic compounds on which life is based, independently suggesting the abundance of the raw materials of life.

And finally, the very molecules of which we are made are found in huge clouds in the inhospitable depths of space. Indeed, the Milky Way is like a cosmic supermarket, filled with the ingredients for the recipe of life.

"I see your little, petrified skull . . . labeled and resting on a shelf somewhere."

Planet Trek

HOW TO FIND PLANETS

SCIENTIST FINDS PROOF OF PLANT LIFE ON STAR
The strongest evidence yet that plants surround a nearby star has renewed speculation about
possible "life on other worlds", says an astronomer who took the first photograph of what may
be a young solar system . . . the first direct evidence that some of the material around one of
those stars—Beta Pictoris—has coalesced to create plants.

SUCH EXCITEMENT from a simple mistake! The word "planet" was misprinted as "plant," and a whole article on the discovery of life thus sprouted from the *Aberdeen Press and Journal* on October 17, 1984.

What had really happened was remarkable in itself, but a long way from the discovery of life. It was the discovery of hints that planets may be forming near another star.

To estimate the chances of life elsewhere, SETI scientists need to know whether planets themselves are common or rare. For a long time, the most intriguing clue we had to the existence of planets around other stars was very indirect. It started with the discovery that stars spin. How do we know? Our first proof was from sunspots. Those dark spots—slightly cooler parts of the Sun's skin—move around the Sun slowly, taking about a month to completely go around and come back. So the Sun spins slowly, like a gigantic planet.

Astronomers looked for signs of spinning stars elsewhere. By studying the spectral lines of other stars, we found they were altered by the Doppler shift. One edge of a typical star is slightly redder than normal, meaning that it's moving away from us, while the other edge looks slightly bluer, because it's moving toward us. We can then calculate how fast it's spinning.

We also have pretty good theories of how stars work: how they generate heat by thermonuclear reactions similar to those of a hydrogen bomb, as well as how they are born, age and die. When we combine these theories with the measurements of starlight, we find a remarkable relationship: young stars (which tend to be hot and bright) usually spin rapidly, dozens of times faster than our Sun. But middle-aged stars (like the Sun) tend to be slow. Stars spin rapidly when young, but slow down as they age.

A spinning star has an enormous rotational momentum, called *angular momentum*. But if you look at our solar system as a whole, you find that most of the angular momentum is *not* in the Sun—it is in the planets. While the Sun has 99.9 percent of the mass of the solar system, it has only a tiny part—half a percent—of the angular momentum. Most of the angular momentum is in the planets, especially Jupiter, due to their movement around the Sun. This suggests that, somehow, the Sun started out spinning rapidly, but as planets formed, they took away most of the spin, perhaps aided by the gas and magnetic fields shed by the early Sun.

Because stars like the Sun usually spin slowly, we suspect planets may be orbiting them.

Wobble Watching

In the last several years, increasingly strong evidence has mounted to support the idea that planets may be common in our galaxy.

Many people think that scientists have discovered planets around

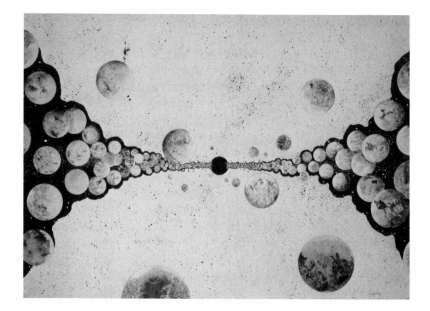

An artist's conception of the possibility of vast numbers of Earthlike worlds.

other stars. Every now and then, articles appear with headlines like *Scientist Finds New Planet.* Not true. It's better than saying they have found plants, but it is still an exaggeration.

What has happened is that scientists have found many *hints* that planets are common in our Milky Way galaxy, but there is not one absolutely, positively, generally accepted discovery of a conventional planet outside the solar system. It is very likely, though, that in the next several years there will be such a discovery.

How do you look for planets around other stars? It is a very tough job. Stars are already trillions of miles away and more. And a planet only glows by reflecting the light of its star, so it would be washed out by the vastly brighter light of its sun.

A remarkable fact is that most stars you see at night are multiple: two, three and even more stars slowly dancing around each other in intricate orbits almost like planets around a sun. Many of these multiple stars can be easily distinguished when viewed with a telescope. Occasionally the stars of a binary pair are so close that the light from one is drowned out by the other. We can sometimes detect the fainter star by watching the brighter one over a period of years. The two stars pull on each other with their gravitational forces. When a small, faint star moves around a big, bright one, they orbit around the center of mass, which is nearer the big star. From far away, it looks as if the big one is bouncing back and forth in space over a period of years.

In the nineteenth century, the great German astronomer Friedrich Bessel was the first to apply this technique to the stars. Bessel, who made precise measurements of the positions of fifty *thousand* stars, noticed a faint wobble in the bright star Sirius. He deduced the existence of an invisible star, called Sirius B. It was not until after his death that telescopes powerful enough to actually see Sirius B were built.

Planets cause similar, but smaller, wobbles. Even the Earth causes the Sun to wobble. But since our planet is 1/300,000th the mass of the Sun, the center of mass of the Sun-Earth system is 1/300,000th of the distance between the centers of the two bodies—well inside the Sun.

Actually, the planets do not strictly revolve around the Sun. They really revolve around the center of mass of the whole solar system. In practice, the most influential planet is the most massive one, Jupiter, which is three hundred times heavier than Earth. It is so massive and so far away that the center of mass of the Sun-Jupiter pair is just outside the surface of the Sun. If extraterrestrial astronomers observe our Sun, they should see it wobble very slightly during the twelve years it takes Jupiter to complete an orbit. From the measurements, they would even be able to calculate Jupiter's mass. But to detect this wobble, they'd need to be able to measure a 700,000-km (440,000-mile) change in the Sun's position—astronomically, that's tiny.

Nevertheless, human astronomers have searched for stellar wobbles, hoping to find planets this way. On several occasions, they thought they had found Jupiter-sized planets. In each case, however, other astronomers could not confirm the finding. The measurements are very difficult and error is all too easy. The mere expansion and contraction of the glass used in photographic plates used years apart can cause significant errors.

The Orbiting Heat-Seeker

Below, left: IRAS, the Infrared Astronomical Satellite.

Below, right: Ring around a star. This is an artist's conception of the ring of comets, asteroids or dust around the star Vega, discovered by the IRAS spacecraft. This is probably something like what the solar system looked like when it was forming almost five billion years ago. Many similar dust rings have been found around other stars.

A very different approach to finding planets is made possible by rockets. Infrared light—invisible heat energy—has a hard time passing through our atmosphere, which absorbs it. In 1983, a satellite was launched to look at the infrared sky above the atmosphere. IRAS (the Infrared Astronomical Satellite, designed by the Netherlands, England and the United States) was the first major infrared telescope in space, or at least the first one not designed to look for enemies. It was spectacularly successful, running longer than ex-

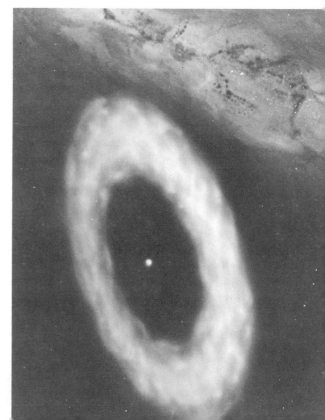

pected and returning thousands of photographs of stars, gas and dust in our Milky Way, revealing many celestial objects never before seen.

From the point of view of planet hunters, the most extraordinary discovery was found around the star Vega. Vega is a young, bright star only twenty-six light-years away, known since ancient times and thoroughly studied by astronomers for centuries. To our surprise, though, something that had never been hinted at from Earthbound telescopes appeared in the IRAS images: Vega was surrounded by a ring of dark matter. Apparently the ring is some combination of gas, dust, comets and other debris that resembles what we think the solar system looked like when the planets first started to form. Planets may be forming there at this very moment.

The photographs taken by IRAS showed that many other stars had similar rings. The star Beta Pictoris had such a ring, which astronomers looked at using a new, ultrasensitive Earth-based telescope system designed to minimize the glare from the star. Bradford Smith of the University of Arizona and Richard Terrile of the Jet Propulsion Laboratory (JPL) were able to take the first pictures of this ring in visible light. This was the discovery that led to the comical article claiming that scientists had observed *plants* forming around stars.

These findings increase our suspicion that planet formation is a normal process of star formation.

Brown Dwarfs and Big Planets

The closest thing to a planet yet detected around another star was found using a different technique. Astronomer Donald McCarthy of Steward Observatory, Tucson, Arizona, took ten thousand infrared photographs of a star called Van Biesbroek 8 and compared it with a nearby star. He found the first *brown dwarf* ever detected—an object somewhere between a planet and a star.

Imagine a giant planet, forty times the mass of Jupiter, crushed into a ball about the size of Jupiter. If you stood on its surface, you would feel a gravity of about 100 G's (you would weigh 100 times what you do on Earth). This world glows deep red like a furnace, and is 2,000 degrees Fahrenheit hot—hot enough to melt gold. This is what we suspect a brown dwarf is like.

Several new searches for planets of other stars are under way. Astronomers like George Gatewood of the University of Pittsburgh now systematically search for planetary wobbles of stars. NASA's Hubble Space Telescope will look for planets part time. The first true planet beyond our solar system will probably be found in the next several years. And NASA's Space Station, if it survives the budget wars of Congress, may be the best place to find planets in the long run.

Planets seem to be a normal byproduct of the evolution of the universe, so the number of places on which life could evolve may be similar to the number of stars. SETI looks ever more promising.

Buck Rogers SETI

EXPLORING THE SOLAR SYSTEM

*Damn the Solar System. Bad light; planets too distant; pestered with comets; feeble contrivance;
could make a better myself.*
Lord Francis Jeffery

LOOKING FOR PLANETS around other stars is important to SETI, but
what about our own solar system? Spacecraft give us close-ups of
our neighboring planets and let us search for life directly.

Sputnik's grandchildren are now scattered around many of the
planets—spacecraft orbit Mars and Venus; others have flown by

The Space Flight Operations Facility at JPL, the control room for
most NASA planetary exploration.

77

Mercury, Jupiter, Saturn and Uranus. They've told us a great deal about the prospects for life elsewhere, and have even hinted at the possibility that it could exist on some of these worlds.

Let's start at the closest planet to the Sun and work outward.

Mercury the Hellish

Hot place. No air. No water. Not a place to look for life, if we are even remotely correct in our understanding of the conditions necessary for it.

Prior to 1974, we knew little about the planet. We knew that it was small—a third the diameter of the Earth—and so close to the Sun that it was hard to observe. Our best effort prior to then had been to "touch" it with radar by transmitting radio signals to the planet and picking up the reflection with an extremely sensitive antenna, just as airports bounce their radar signals off airplanes. Radar showed that Mercury spins very slowly, taking two-thirds of its year to complete one rotation.

Then, in 1974, NASA's *Mariner Venus Mercury* spacecraft flew by Mercury and took its first close-up photos ever. It looked remarkably like the Moon, heavily cratered and airless. This gave important clues about the formation of the solar system: the cratering is a record of the bombardment that the planets experienced during those early, violent days.

But life seems ruled out there.

Venus the Smoggy

Venus: A beautiful, mysterious, cloud-enshrouded world, slightly smaller than Earth and sometimes called Earth's twin.

Before spacecraft, not much was known about it, but Venus drew spaceprobes much as the legendary goddess drew lovers. Soviet and American spacecraft flew by and orbited it, photographing the clouds and using radar to scan the surface.

In 1985, in a mission involving the Soviet Union, France and the United States, the two Russian *VEGA* spacecraft reached Venus, dropped their probes and balloons, and flew on to explore Halley's Comet. On their monitors, scientists at NASA's Jet Propulsion Laboratory in California watched the Soviet data come in from Venus. The drama unfolded millions of miles away, as the balloons were released from the spacecraft and bobbed about in the thick, hot atmosphere.

The mission was a success. The *VEGA* balloons taught us far more about the weather on Venus than any previous mission.

Between the spacecraft visits and Earth-based astronomy, we now know that Venus' atmosphere is mainly carbon dioxide, almost a hundred times the pressure of the Earth's air and hotter than a kitchen oven. The detailed composition of the atmosphere tells us much about the evolution of planetary atmospheres in general, which is improving our understanding of our own planet's atmospheric history.

There is no significant amount of water. The lovely clouds are largely sulphuric acid. The goddess of love is now the planet of smog.

Many mysteries remain. In 1988, NASA plans to send the *Magellan* spacecraft there to make radar pictures as good as photographs of the hidden surface. The French and Soviets are considering sending more powerful balloon-lander probes to Venus in 1989.

The conditions for life as we know it do not exist there now. But one day in the distant future, humans may bring water there, perhaps importing it from comets. We may use bacteria or plants to turn the carbon dioxide into oxygen. One day, Venus may be a liveable world worthy again of its beautiful name.

Below, left: The Moonlike planet Mercury, as seen by the *Mariner 10* spacecraft.

Below, right: Venus, as seen in ultraviolet light by the *Mariner 10* spacecraft. In ordinary light, Venus is virtually featureless.

Earth the Lively

Earth is the next planet out. We know there is intelligent life there, but not very.

Planet Earth, from the *Apollo 10* spacecraft.

Mars the Intriguing

I will never forget the excitement when we landed on Mars. I then worked at JPL, NASA's headquarters for planetary exploration in Pasadena, California. The first picture came back from the *Viking* lander on July 20, 1976. Line by line, the image formed on the screen, gradually revealing the first close-up picture ever taken of that planet. The first image was in black and white—color would come later, after more photography and computer processing.

The image from Mars was incredibly sharp, better than the six o'clock news on my TV. Bit by bit, the Martian landscape was revealed. Rocks and sand and hills and valleys. It wasn't Percival Lowell's or Ray Bradbury's Mars. There were no canals. But, secretly, I hoped that the outline of a Martian or a lizard or at least of some otherworldly plant would be revealed. Alas, that did not happen. But we learned a tremendous amount about Mars.

The two *Viking* landers went to different spots. They were really robots, with on-board computers for brains, giving them roughly the I.Q. of a grasshopper. Although fixed in place, they had sample arms that reached out to different spots near the landers, bit off pieces of Mars, put them in their "mouths," which fed their "stomachs"—miniaturized laboratories. At first, it seemed that two of the experiments had detected life.

Eventually, however, it appeared that they had been fooled by

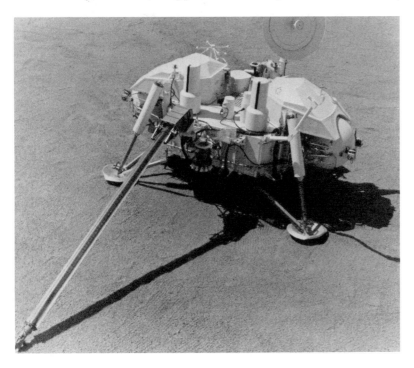

The *Viking* lander, photographed before its voyage to Mars.

strange chemical reactions triggred by the intense solar ultraviolet light that penetrates the thin atmosphere. Today, one scientist thinks they detected bacteria, but most disagree.

Although Mars has only half the diameter of the Earth, it is remarkably similar to our planet in several ways. This is what the picture was when the *Viking* project was being designed. "Put yourself in, say, 1960-63," said Bruce Murray, former head of JPL, "and you're informed that Mars has the same obliquity [tilt] as the Earth, therefore the same seasons. It has white [polar] caps that go from two hemispheres, back and forth. It has the same rotation

Mars, viewed from a *Viking* orbiter. The huge Grand Canyon of Mars, Valles Marineris, stretches across most of the width of this image. (Earth's Grand Canyon would be lost inside it.) The dark sides of three huge volcanoes are visible at left.

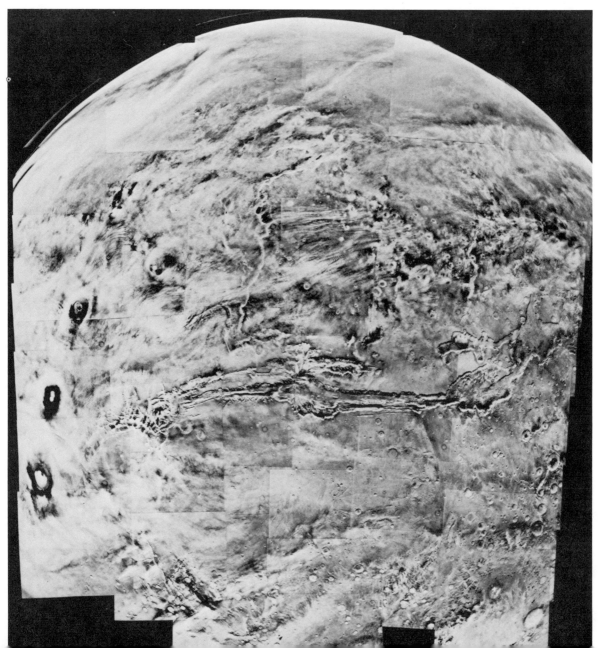

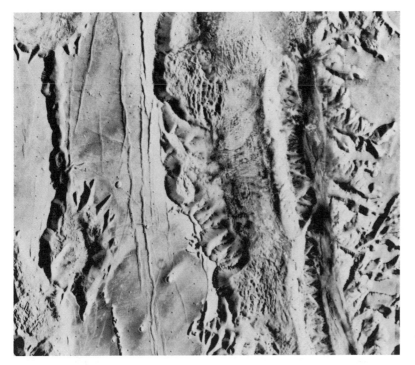

The Valles Marineris canyon, close up. Some of its features may have been formed by water.

period within thirty-five minutes of the Earth, and it has variations in light-and-dark markings that change and migrate seasonally, that start at the equator and move out to the poles, just like Earth."

Percival Lowell, back in the nineteenth century, had speculated that these seasonal changes were due to plants growing on Mars, and in the early 1960s that theory was still respectable.

"And so to say that it was certainly the most Earth-like planet was a very useful assumption. In addition, you had this hard, physical evidence from '56 and '60, by a *very* respected astronomer, that the dark areas on Mars had some infrared absorptions that correspond closely to organic chemicals." This astronomer had found, using an Earth-based telescope, that the light from Mars seemed to contain spectral lines of a chemical that is found in living things on Earth.

So to scientists back then, it seemed that Mars was the place to look for life. Unfortunately, much of their information was wrong.

We now know, according to Murray, that although Mars is tilted by 24 degrees—about the same amount that gives Earth its seasons—this tilt actually oscillates between 15 and 35 degrees, every two million years. It just happens currently to be at the point in its cycle where it is tilted the same as the Earth.

And the white substance that was observed to move back and forth

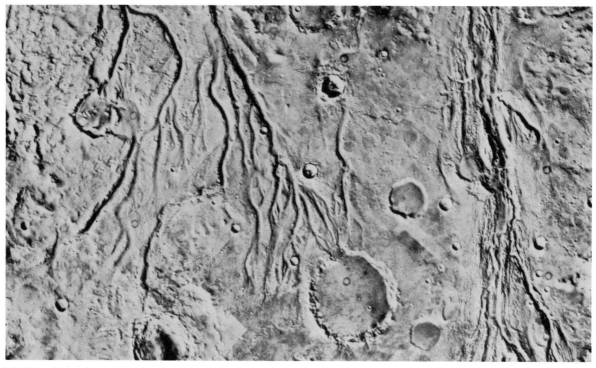

Winding channels on Mars, probably caused by a vast flood in ancient times, one of the pieces of evidence that the planet was once very wet.

seasonally between the two poles, which was thought to be water frost, is actually frozen carbon dioxide. There is water frost, too, but it's not nearly as abundant as it seemed back then, although there is much evidence for ice underneath the surface of the planet. Furthermore, the light-and-dark markings that were thought to be evidence of plants, we now know are wind-blown dust.

But what about that detection of organic chemicals on Mars, the seemingly ironclad proof of life? It turned out to be caused by a rare form of water vapor in the Earth's atmosphere. It had nothing to do with Mars at all. A fine astronomer had made a mistake, understandable in the wake of the Lowell legacy, which in itself made the search for life on Mars the central focus of the U. S. planetary program.

In 1965, *Mariner 4* made the first successful flyby of Mars and found huge craters there. In fact, Mars looked so Moonlike in these early, low-quality images that some scientists went overboard in the opposite direction and concluded that life must not exist there. Fortunately, Caltech biologist Norman Horowitz designed the crucial *Viking* lander experiment that later tested for life on Mars, and he was not misled by either the pro-life or the anti-life extremes. Instead, he learned from the geologists that the lack of erosion in the craters proved there could not have been any rainfall in a very

Giant tubeworms growing at the bottom of the oceans thousands of feet down, near natural hot-water vents. In the last decade, such lifeforms have been found to be based on a radically different ecology and chemistry from all other life on Earth. This means that the conditions under which life can arise are even more diverse than had been previously thought.

An infrared view of the Andromeda galaxy, a swirl of billions of stars similar to our Milky Way. Taken by the IRAS spacecraft, the computer image shows bright areas where stars appear to be forming from clouds of gas and dust.

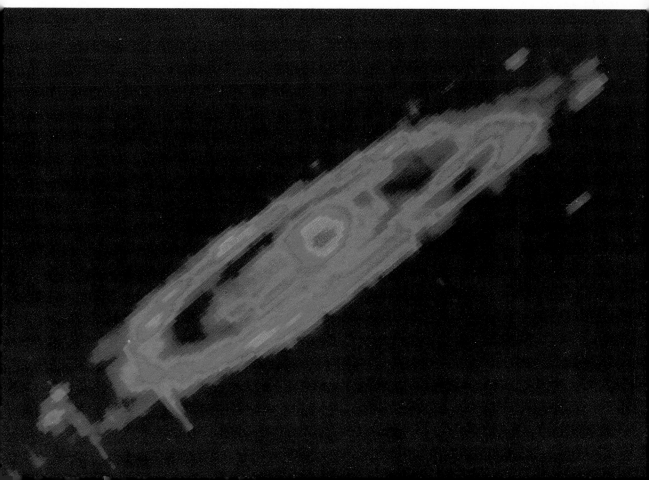

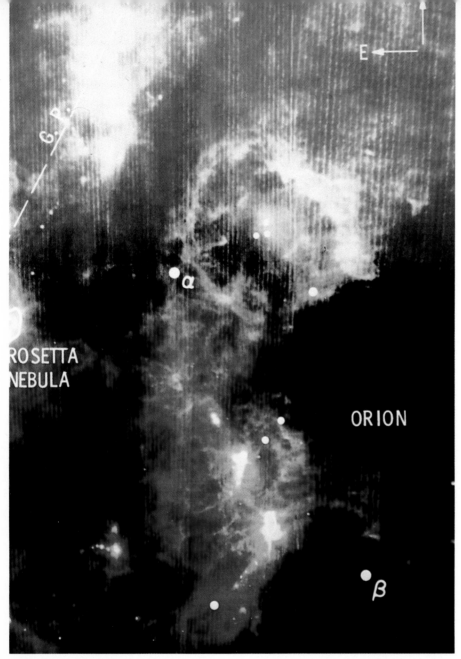

The familiar constellation Orion, as seen by the IRAS infrared space-telescope. The glow is from gas and dust, the materials out of which molecules and stars form.

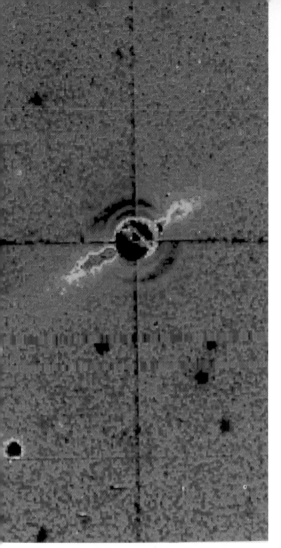

Planets being born? This is the star Beta Pictoris, 50 light-years away. Photographed from Earth with a mask blocking off the light of the star, it shows orbiting material that may be starting to form planets. The dark straight and circular lines are not real.

The best photograph of Halley's Comet taken by the spacecraft that flew by it in 1986, from the West European Giotto probe.

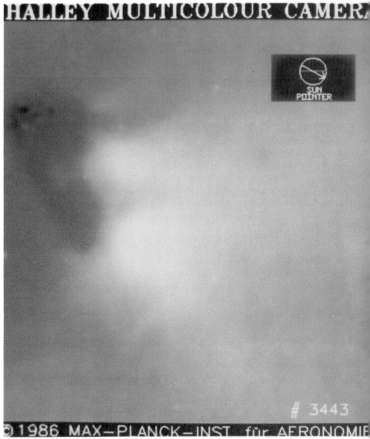

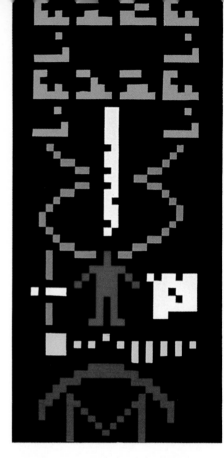

Above, left: The 1000-foot diameter radiotelescope in Arecibo, Puerto Rico, where the most sensitive SETI work has been conducted.

Above, right: The Arecibo Message, an exercise in sending the most information in a short signal. It was sent in 1974 as a binary string of zeroes and ones that could be assembled to make a simple picture, like a TV image. The image used a modified binary numbering system, with an extra bit (dot) added to each number to make it clear where each number starts. At the top is the count from one to ten, then a block represents the atomic weights of the five most common atoms of our life (hydrogen, carbon, nitrogen, oxygen and phosphorous), side by side. Then there are formulas for basic molecules and radicals of life, expressed in terms of the numbers of those five atoms. The double helix of the DNA is shown, sandwiching the number of molecular base-pairs in human DNA (4 billion). Under that is a crude picture of us, between a measure of the average human height in units of the signal's wavelength (5 inches), and the population of the Earth. Under the person is a picture of the solar system, with Earth displaced toward the body, indicating where we live. Finally, there is a cross section of the radiotelescope, with a picture of the radio signal bouncing out, and a measure of the antenna's size.

   With more time and money, such as an alien civilization might have, this message could be made much clearer.

Right: Project META's 84-foot diameter radiotelescope at Harvard, Massachusetts. Horowitz' young son stands at the base.

long time. So he designed a biology experiment for the *Viking* landers that did not use water.

Horowitz' experiment exposed a sample of Martian soil to carbon dioxide and carbon monoxide. These gases were labeled with radioactive carbon 14, which could be easily detected, just as doctors sometimes use small amounts of radioactive chemicals to trace the flow of medicine in the body. His device then tested whether the Martian sample "breathed" the gas as any self-respecting carbon-based lifeform such as an Earth plant would do.

His experiment was based upon the cycle of cooking the sample and looking to see whether there was any biological interaction with carbon dioxide. The other experiments that were aboard essentially threw water onto the sample. They put the soil into water, which was mixed with nutrients, and it *exploded*, as if the surface of Mars had never seen water, as if it were "allergic" to water.

So the *Viking* landers, sadly, did not find life.

Spaceprobes have given us profoundly important information about the geology and atmospheres of our two nearest planetary neighbors, Mars and Venus. At last, we have data with which to test our theories of the history of our own planet. Mars and Venus have

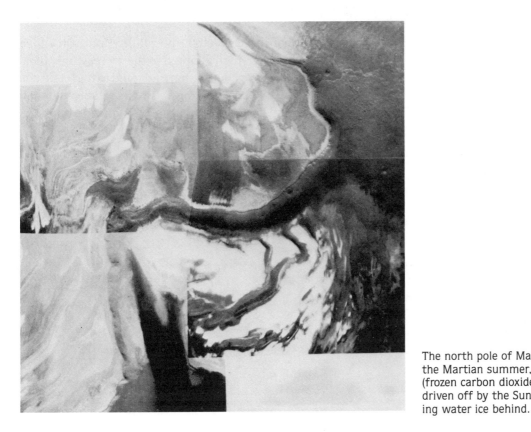

The north pole of Mars. This was the Martian summer, and dry ice (frozen carbon dioxide) has been driven off by the Sun's heat, leaving water ice behind.

shown us many surprises with their strange geological features and puzzling details of atmospheric gases. They've forced us to revise our theories of our own planet's history. With Venus on one side of us and Mars on the other, we are beginning to calibrate the Earth, to learn where our planet fits in among the possibilities of the universe. And this gives us a better grasp of the possibilities of life on other worlds.

It's important to remember, however, that so far we have only been to two different spots on Mars. Those spots were chosen to be "safe," not to be pleasant for life. The Russians had previously landed a probe on Mars, but it failed as soon as it started to transmit. They joked that a Martian had kicked it over. The truth was probably that it had landed on the edge of a crater or boulder and slid, its antenna unable to point to Earth. Knowing this, NASA's mission planners made sure the *Vikings* landed in smooth places.

They were not near riverbeds or ice caps, nor near geological features that seem to be caused by underground ice, where water might conceivably be available to some form of life. Most scientists think that *Viking* ruled out any possibility of Martian life. However, there is always the possibility that life could have evolved billions of years ago, when Mars was wet, when there were great rivers and a thicker atmosphere.

It seems to me that there's a slight possibility that some form of life could have evolved which adapted to the increasingly bleak conditions on Mars, just as penguins have adapted to the frigid conditions of Antarctica. Even if just a lowly fungus were found there one day, it would be an enormous breakthrough in understanding the possibility of life on other worlds. And even if there is nothing living on Mars today, perhaps we will one day find fossils of past life there.

A close-up of the Martian surface from the *Viking 2* lander. Many of the rocks looked quite different from those at the first site, being filled with holes like some types of lava rocks on Earth.

Could there once have been Martian dinosaurs roaming the vast canyons of Mars? Mars was probably not wet long enough to evolve advanced animals, but there might have been some primitive form of life. The detection even of fossils of the most primitive cells would be a tremendous step forward in our understanding of the evolution of extraterrestrial life. It would prove that life could evolve independently on two different worlds. If that were true, the likelihood of life on worlds of distant stars would be vastly increased and even the critics of SETI would have to rethink their objections.

In 1988, the Soviets will send a spacecraft to study the Martian moon Phobos, complete with a miniature "Star Wars" probe to measure the composition of the body. The probe will use a laser or particle beam to vaporize tiny parts of Phobos to learn its composition. Two years later, NASA will send a Mars Orbiter to study the geology and climate of the planet.

One day, NASA hopes to send a roving robot explorer to Mars. It would land at a safe spot and travel to the more interesting ones.

Beyond that, there is a movement in the United States and the Soviet Union to send astronauts to Mars. In preliminary discussions between the two nations, scientists have explored the possibility of a joint mission. They have even considered establishing a permanent base there, much like our Antarctic ones.

The Martian atmosphere, like Venus', is composed largely of carbon dioxide, and can be broken down into oxygen and carbon monoxide. The oxygen we could breathe; the carbon monoxide could be used as an unconventional rocket fuel when burned with oxygen. We already know that water, essential for life as well as much industry, is present in the ice caps. And there is recent evidence, from a reappraisal of *Viking* orbiter photographs, that ice may be available in warmer areas under the surface.

Mars is so Earthlike compared to the other planets that I am confident that it will one day be the first planet beyond ours that humans call home.

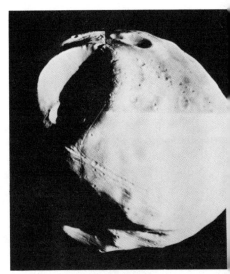

The Martian moon Phobos, possibly a captured asteroid from the belt that lies between Mars and Jupiter. This object is only about 13 miles in diameter.

Asteroids the Mountainous

Between Mars and Jupiter lies the great asteroid belt, mountains of rock, some of them a fifth the diameter of the Moon. Many meteorites that hit Earth probably originate in the asteroid belt. (The rest probably come from comets.) They are too small to hold an atmosphere, but are still of indirect interest to the search for life.

Asteroids form a belt around the Sun much like the rings of

Saturn, though much more spread out. They seem to be the remains of small Moonsize objects that collided eons ago.

The most fascinating aspect of asteroids is that some meteorites on Earth—probably ex-asteroids—contain complex organic compounds, the building blocks of life. There may be huge mountains of such chemicals orbiting the Sun, and they may have helped trigger the formation of life here. That such chemicals formed in space (both in the asteroids and in interstellar clouds) is one more piece of evidence that the essentials for life are readily available in the universe.

Jupiter the Giant

It has been said that the solar system consists of the Sun, Jupiter and other stuff. By far the biggest planet in our system—eleven times the diameter of Earth and three hundred times as massive—Jupiter

The Earth and Jupiter, the largest planet, to the same scale.

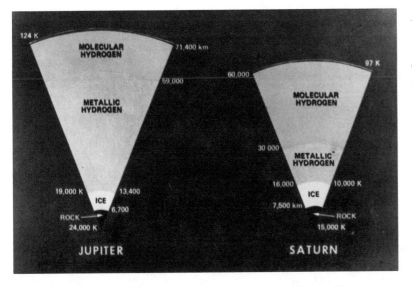

The interiors of Jupiter and Saturn, as we think they are. Both planets are made mostly of hydrogen, which is a gas in the outermost layer, and a liquid below that. Deep inside, the liquid is compressed so much that it becomes a metal able to conduct electricity. This is probably where powerful electric currents flow that generate the planet's magnetic fields. (Temperatures are in degrees Kelvin [K]—degrees centigrade above absolute zero. For comparison, room temperature on Earth is about 300 K.)

has an enormous, permanent storm in its atmosphere, called the Great Red Spot, large enough to hold three Earths side by side.

Four spacecraft have flown by Jupiter: *Pioneers 10* and *11*, and *Voyagers 1* and *2*. The *Pioneers* took the first pictures; the *Voyagers* had far more sophisticated equipment and took the exquisite photographs that graced most of the front pages of the world's newspapers and magazines.

Jupiter's atmosphere is mainly hydrogen, helium, ammonia and methane, just what the Earth's primitive atmosphere probably contained. (The helium, being light and chemically unreactive, escaped.) It even has huge thunderstorms to provide electricity.

Could life exist there?

Most scientists would say no, but a few remain cautiously optimistic. The problem is heat. Inside Jupiter, it is very hot, so hot that Jupiter radiates about twice as much heat as it receives from the Sun. (This may be a legacy of the formation of the solar system: Jupiter may still be cooling down.) The deeper you go into its atmosphere, the hotter it gets. Organic chemicals tend to break down when heated, much as your food turns to gunk when left on the stove too long. The orange-cloud color may be from organics.

Still, at the cloudtops—the part you see in a photo of Jupiter— the temperature and pressure are about the same as in the place where you are reading this book. It is conceivable that creatures could have evolved to fly, swim or float in such an environment. The equivalent of birds or jellyfish could live there. Creatures with natural hot-air balloons could thrive. Evolution is so adaptable that we should not rule out the possibility of life there until we have explored it more thoroughly.

In fact, NASA plans to send *Galileo*, the most sophisticated robot spacecraft ever, to Jupiter in the late 1980s. The probe is built and awaits only the availability of the Space Shuttle. One part of *Galileo*, the orbiter, will circle Jupiter while the other part, the probe, actually enters the atmosphere.

The *Galileo* probe will probably survive the horrendous heat of entry for an hour or so. Unfortunately, it has no camera, so we will not see any pictures of Jupiter's cloudtop inhabitants if they exist, but it will make many measurements to solve some of mysteries about what lies below those clouds.

A montage of *Voyager* pictures of Jupiter and its moons: Io (upper left), Europa (center), Ganymede (lower left) and Callisto (lower right).

A surprising prospect for life is in the moons of Jupiter. The planet has four moons as large as our own, which were discovered by Galileo. The outermost three (Europa, Ganymede, Callisto) are icy and airless and, at first glance, unpromising. The innermost of the Galilean moons, Io, is extraordinary.

Io is crazily colored, with splotches of orange, yellow and black, "like a diseased pizza," as one scientist described it. Io is the most volcanic object we have so far discovered in the whole solar system. When the *Voyagers* flew by it, half a dozen volcanoes were erupting. As the spacecraft flew by, the volcanoes spewed lava and gas hundreds of miles into space, forming a thin atmosphere (though still a near-vacuum by our standards). Part of the atmosphere escapes

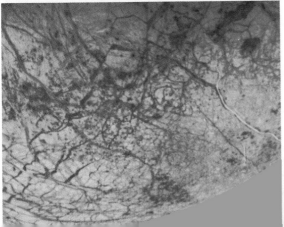

Above, left: Io, one of the weirdest places in the solar system, where half a dozen volcanoes were erupting while the *Voyagers* flew by.

Above, right: Io, showing two of its volcanoes erupting, spewing material hundreds of miles into space. One is on the rim, and the other on the dark side (seen as its plume rises into sunlight near the day-night boundary).

Left: Europa, the next moon out from Io, probably has a liquid ocean underneath its icy surface.

into space, encircling Jupiter with a doughnut of transparent, faintly glowing gases.

Why Io? Just before the *Voyagers* found the volcanoes, scientists had predicted that the constant flexing in Jupiter's titanic gravitational field would heat Io's insides up. This moon is so close, in fact, that its shape is slightly distorted by the pull of Jupiter's gravity, just as our Earth's oceans are pulled by the Moon's gravity, giving us tides.

If there were no other moons around the planet, Io would simply orbit in a circle, permanently squeezed into a pear shape, and nothing exciting would happen. But because the other moons pull on it, tugging it slightly away from a perfectly circular orbit, Io feels a slightly varying gravity, a tide. Io "breathes." And this flexing heats the moon, just as the rapid flexing of a metal strip makes it become warm. Io's interior is therefore molten and gushes forth from the spots we call volcanoes.

Does this help the chances for life? Probably not on Io, because there seems to be no water available. But it suggests that interesting things may be happening inside the next moon, Europa. Although covered by ice, the same mechanism that heats Io heats Europa, but less effectively since it is farther from Jupiter. It is not hot enough to create volcanoes, but it is probably hot enough to melt ice. There seems to be an ocean inside Europa. And once you have an ocean, the possibility of life becomes very real.

Below, left: After Europa comes the frozen moon of Ganymede with bizarre patterns in its ice.

Below, right: Callisto, the most distant of Jupiter's large moons, is heavily cratered, showing an extraordinary bull's eye almost as big as the moon itself. This feature was probably made in the ice by a meteorite.

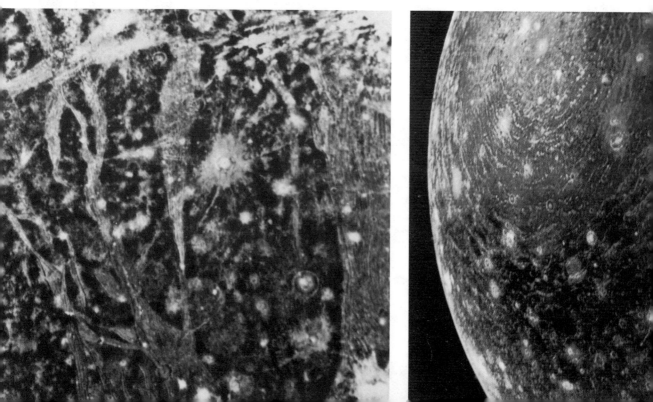

Galileo, the first spacecraft that will drop a probe into Jupiter's atmosphere.

One day in the distant future, I suspect we'll send probes inside Europa's arctic ocean, entering through a giant crevasse and tunnelling through the ice. Perhaps there will even be scuba divers and submarines. Eventually a Jacques Cousteau may swim deep inside Europa, under the dark, eerie ice cover. The diver's lights will stab through the gloom and the explorer will see with his or her own eyes whether there are other creatures swimming there.

Long before then, the *Galileo* orbiter will fly much closer to the large moons than its predecessors did, and send back even more spectacular pictures of these colorful worlds.

Saturn and Earth, to the same scale.

Saturn the Ringed

Saturn's ringed beauty has been renowned since shortly after the discovery of the telescope. It's the second largest planet in the solar system, a little smaller than Jupiter, but still nine times the diameter of the Earth.

Pioneer 11 and the two *Voyagers* revealed many secrets of this planet, confirming our suspicions that in many ways, Saturn is a smaller version of Jupiter. (Jupiter even has a faint ring, discovered by one of the spacecraft.) In the Saturnian atmosphere, the same possibilities of relatively pleasant living conditions (for a hydrogen-helium-ammonia-methane breather) exist as on Jupiter. Also, the same excruciatingly hot conditions exist below the comfortable atmospheric layer.

Among its many moons, Saturn has one, Titan, which is in its way as unusual as Io or Europa. Titan is one of the biggest moons in the solar system, and its claim to fame is that it is the only known satellite with an atmosphere thicker than Earth's. As with Io, part of the atmosphere leaks off into space, forming a giant doughnut of transparent gas surrounding Titan's orbit. The orbit is about ten times the diameter of the rings of Saturn, so if you could see the doughnut, you would see Saturn encircled by rings which are in turn encircled by a doughnut larger than the Sun.

We've known since World War II that Titan has an atmosphere, but we did not know until *Voyager* flew by it how extraordinarily thick the atmosphere is. Astronomers were excited at the prospect of seeing Titan's surface for the first time, but became rapidly disappointed as the spacecraft approached the moon, because they couldn't see anything except a bland cloud-cover.

The astronomers' disappointment with Titan turned out to be

Below, left: Saturn's moon Dione. The wavy valleys may be caused by geological faults.

Below, right: An artist's speculative view of the gloomy surface of Titan. A rain covers the ground with organic chemicals and a glacier winds through a valley.

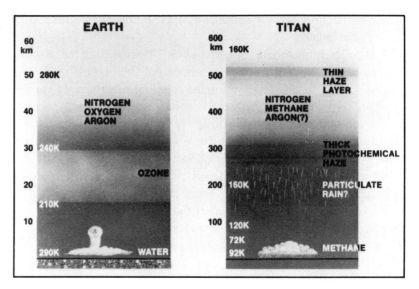

The atmospheres of Earth and Titan, compared. Both worlds have atmospheres primarily of nitrogen, but instead of oxygen, Titan has methane and other gases, and may have snows and seas of exotic chemicals, including organic compounds.

a blessing in disguise. It meant that its atmospheric chemistry is much more complex, and hence much more interesting, than is possible with a thin gas. The atmosphere is mostly nitrogen, with methane, ammonia and related compounds, but oceans of organic chemicals may exist there. Also, the reddish-brown haze appears to consist of organic chemicals. While "organic" in this sense means basic carbon compounds, once again there is a possibility that living things could have evolved there, though most scientists think the present temperatures are too cold for life. Studies are under way in Western Europe and the United States to design a *Galileo*-like probe called *Cassini* that would orbit Saturn and probe Titan.

Uranus the Tipsy

Uranus, the next planet, is also a giant, four times the diameter of Earth but only one-third that of Jupiter.

Uranus was one of the most mysterious planets until *Voyager 2* flew by in 1986. Launched in 1977, this spacecraft had fulfilled its mission when it returned beautiful pictures and other data from Saturn in 1981. Scientists prayed that it would last as far as Uranus, but for a while it looked as if it would not. The scan platform that aims the TV camera had jammed at Saturn; the main receiver had failed long before; the backup receiver had deteriorated and its frequency drifted; various electronic chips failed on the way to Uranus. *Voyager* was old, arthritic and senile. To make matters worse, Uranus is twice as far from the Sun as Saturn, making the light four times dimmer (or 400 times dimmer than when viewed from Earth), and making the spacecraft's signals proportionately weaker.

Uranus conspired in another way to make this probably the most difficult planetary encounter ever. We have long known, from murky photographs of the dim planet taken through Earth's atmosphere, that Uranus has a most peculiar feature. It is tilted "sideways."

Normally, a planet spins much like Earth, with its axis of rotation roughly perpendicular to its orbit, like a toy top spinning on a table. But Uranus's rotation axis lies roughly in its orbital plane. It is as if the Earth were tilted over so that the South Pole faced the Sun. If this were the case, the South Pole would melt completely and the North Pole would become even colder than normal. Then, three months later (a quarter of an orbit), the Sun would be over the equator, just as it normally is. But in another three months, the North Pole would face the Sun. This time, it would be the North Pole that would melt while the South Pole froze.

The best Earth-based picture of Uranus, showing its rings and its five large moons. Richard J. Terrile (JPL) and Bradford A. Smith (University of Arizona) used a super-sensitive camera called a charge-coupled device (CCD) at the Carnegie Institution's Las Campanas Observatory in Chile. (The vertical lines are noise.)

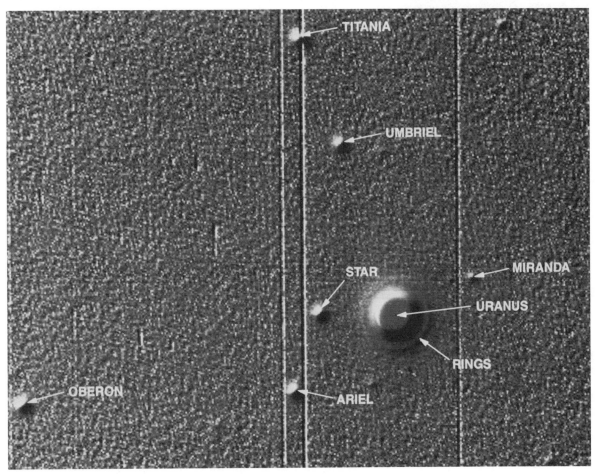

This is essentially what happens to Uranus, only it is so far away that it takes eighty-four years to orbit around the Sun completely, so its seasons last twenty-one years each. By chance, Uranus happens to have its South Pole facing the Sun right now, so *Voyager* gave us the first look ever at such a peculiarly oriented world. The weather patterns must be strange, radically different from those of any other world.

Voyager gave us good news and bad news. The bad news is that the cold outer atmosphere has a thick cloud layer, preventing us from seeing very much. Uranus is fogged in.

The good news is that the moons of Uranus are fascinating. Astronomers, knowing they were very cold, expected them to be dull ice-worlds with little signs of geological activity. Instead, they found that these moons are covered with all kinds of markings indicating a geologically exciting past.

Miranda, the closest of the major moons, is the most curious. It's a crazy quilt of geological faults, streaks, canyons and other structures that make it unlike any other moon yet seen.

Perhaps the most important outcome of the Uranus encounter was that it has added to our understanding of the history of the solar system. The craters on the moons indicate, as on many other planets and moons, that the early history was one of titanic collisions. Giant meteorites crashed into planets with energies dwarfing that of all the bombs on Earth.

Voyager also helped us understand why Uranus is tilted on its side, and that too is a clue to our planetary history. One theory of the tilt is that Uranus was hit by a large moon, which knocked it over. Scientists expected the planet's magnetic field to be similarly tilted, but in fact *Voyager* found it was tilted substantially less. By studying the magnetic data, we'll get important clues to the invisible interior of Uranus, which may help unravel the history of the planet.

The rings of Uranus provide another clue to the past, a more subtle clue that may be most important in the long run. Its rings are similar to those of Saturn but are much thinner and darker, so much so as to be extremely hard to photograph from Earth. They are among the darkest substances in the solar system—as black as charcoal. They are made of dark boulders, perhaps of methane ice, orbiting Uranus in complex ways that may mimic the way planets formed around the Sun.

Because of *Voyager*'s observations of the rings of Saturn and Uranus, and the even fainter ring of Jupiter, we are beginning to understand how such rings behave. Rings may be formed when a small moon collides with another moon. The gravitational fields

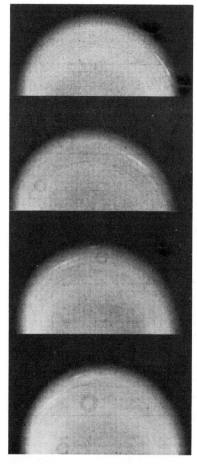

Four *Voyager* pictures of Uranus taken over a five-hour period, showing two bright clouds moving as the planet spins. Few other features were seen in this atmosphere, where cold temperatures keep the planet shrouded in clouds. (Small circles are from dust on the camera lens.)

of the planets and other moons sometimes shepherd the fragments into neat, stable rings. Following the Uranus observations, scientists immediately began to modify their ring theories to take into account conditions of unusual tilt and elongation not seen in other planetary rings. Caltech's Peter Goldreich then suggested that ring particles may stick to one another and eventually form into a moon again over hundreds of millions of years, leading to a wonderful vision of an "ecology" of rings: rings born from moons, and moons reborn, phoenixlike, from the "ashes" of the rings.

It may be that we are seeing in these rings a scaled-down model of the formation of planets.

From the point of view of the search for life, the most important result of the flyby was indirect evidence that there is a hot ocean of water beneath the clouds. Perhaps 5,000 miles deep, this ocean was probably created by comets. Billions of comets orbit around the Sun outside the planets. Uranus and Neptune, having powerful gravitational fields that attract passing objects, and being on the edge of the planetary system, probably have suffered far more collisions with them than have any other planets. Since comets are full of the basic chemicals of life, this also means that one could imagine an ocean of swimming creatures there: Uranian fish and whales. The ocean is very hot, however, with temperatures of thousands of degrees, similar to the heat inside the Earth caused by natural radioactivity. This is probably too hot for life, causing complex molecules to break down before they can evolve into interesting beasts.

Below, left: A composite of *Voyager* pictures of Uranus and its closest moon, Miranda, with an artist's conception of the rings.

Below, right: One of the strangest moons in the solar system: Miranda. It is a hodge-podge of different types of geological features, perhaps more types than any other moon. It may once have blown apart in a collision between moons and been reassembled by gravity.

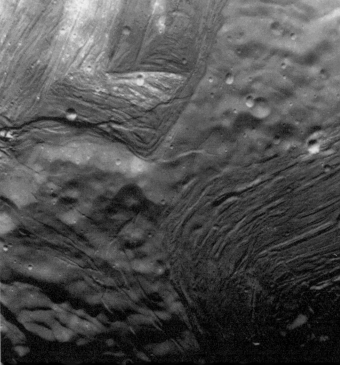

Still, Uranus suggests another radically different way a planet could be suited for life despite its great distance from a star. There may be worlds similar to Uranus but cooler, where comets have brought water and the basic chemicals of life, filling oceans with swimming creatures.

Overall, the *Voyager* flyby was a superb success. The scientists showed their appreciation for the engineers' spectacular job by awarding them two cases of champagne.

It was bitterly ironic that, on the very day the *Voyager* scientists were scheduled to hold their last, triumphant press conference, Space Shuttle *Challenger* exploded after launch. The mood at JPL changed that day from joy and excitement to unrelieved gloom.

I doubt that I shall ever forget the haunting scene of the NASA television monitors at JPL, silently showing, hour after hour, nothing but the ocean by Cape Canaveral, where the Shuttle had gone down. Pieces of Shuttle debris bobbed gently on the waves . . .

Neptune and Beyond

Neptune is much like Uranus, but even farther out. It's the last of the giant planets, and nearly the same size as Uranus. Less is known about it than was known about Uranus prior to the *Voyager* flyby. *Voyager* will fly by Neptune in 1989, and if it survives that long, will give us the first close-up pictures ever of that world and its big moon, Triton.

Triton is a mysterious world about the size of the planet Mercury. Not much is known about it except that it has an atmosphere of

A computer simulation of *Voyager* flying by Neptune and its major moon, Triton, in 1989.

nitrogen and methane, about a tenth the pressure of Earth's (around ten times thicker than Mars'). It may have lakes or oceans of liquid nitrogen, and some type of organic slush may conceivably exist there.

Pluto, whose orbit marks the edge of our planetary system, is only a fifth the diameter of the Earth, has one moon and a very thin methane atmosphere. We have never seen it clearly, and no spacecraft is scheduled to go near it, but the planned Hubble Space Telescope may show features from Earth orbit which cannot be seen from the ground. Until then, the best we can say is that Pluto is so far away and so cold that it seems quite impossible for life as we know it.

There are intriguing hints that there may be a planet beyond Pluto. There seem to be subtle disturbances in the motion of Neptune which may be caused by a distant planet. So far, no such planet has been found, but astronomers keep looking.

Beyond Pluto, a billion billion comets probably fly. Most orbit around the Sun well beyond the orbit of Pluto. Comets such as Halley's, with orbits taking them inside the planetary system, are quite rare. Frigid comets, far from the stimulating energy of the Sun, are not encouraging places in which to look for life, but they contain life's main ingredients.

Comets consist of ices of water, carbon dioxide (dry ice), methane and ammonia, very much like the atmosphere of early Earth. At the very least, they are crucial clues to the origin of life, because they evidently formed when the planets did, but contain chemicals that have long since disappeared or have been altered on the planets. Comets are the frozen leftovers in the cosmic refrigerator.

Comets may even have triggered the origin of life. Two of the controversial scientists we met earlier, Sir Fred Hoyle and Chandra Wickramasinghe, have even gone so far as to suggest that comets contain frozen viruses or bacteria which could have given rise to life and, they claim, triggered plagues and other diseases on Earth.

Most scientists disagree with this extreme view, but it is not unreasonable to suspect that comets may at least have brought some of the building blocks of life to Earth.

Based on our exploration of the solar system thus far, most experts think life does not exist on these other planets. I am not quite so willing to give up hope. There are still so many unknowns in the story of life's origins that it would be premature for us to rule it out completely. Astronomy and space exploration have made too many unexpected discoveries to be confident of such a conclusion: the insides of the giant planets are hot, not cold; the moons of the outer planets are strange little worlds, with active volcanoes on Io, substantial atmospheres on Titan and Triton, possibly an ocean inside Europa.

On Earth, our experience with chemistry has been highly biased. We were born on a lukewarm, watery planet with an oxygen atmosphere and carbon-based life. Most of our chemistry has dealt with compounds formed under conditions not terribly different from what we are used to. On other worlds, with different atmospheres, different fluids, different soils, different temperatures, there may be whole realms of complex chemistry which could, over billions of years, lead to some form of life.

In fact, a recent development in chemistry suggests that life might indeed be able to evolve in extremely cold temperatures, such as exist on most of the moons of the outer planets, and inside comets and asteroids. Most chemical reactions slow down as the temperature falls, and it has long been assumed in SETI that Earth-like temperatures are needed for life. But the startling world of *quantum theory* hints at other possibilities.

A possible future mission to a comet, using solar cells to convert sunlight into electricity, to power ion-drive engines using electrically charged metal atoms instead of rocket fuel. Since we think comets are the frozen leftovers from the formation of the planets, they probably contain valuable clues to our origins. This mission was originally intended for Halley's Comet, but was never funded.

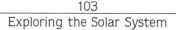

Cut-rate comet explorer. The first in NASA's low-cost *Mariner Mark II* series of spacecraft will probably be this mission to fly by an asteroid or two and rendezvous with a comet, staying with it as the comet approaches the Sun and forms its spectacular tail. The spacecraft, called *CRAF—Comet Rendezvous/Asteroid Flyby*—is shown in this artist's conception flying by an asteroid. Such a mission would return critical clues to the formation of planets and the evolution of atmospheres and life.

Quantum theory is the branch of physics describing the submicroscopic world of atoms and molecules, a world where electrons and protons often act more like waves than particles, where the behavior of the universe often violates our intuition. Thanks to quantum behavior, some chemical reactions that we had expected to stop at low temperatures can actually proceed. This opens up whole new vistas of possible cold evolution in our solar system and elsewhere.

We must also never forget the power of evolution: once life arises through any process, no matter how improbable, it can adapt to changes in the environment. The air of today's Earth would probably have been poisonous to the first life. Geologists have documented numerous mass-extinctions in the past, such as the one in which the dinosaurs disappeared, yet life adapted and continued.

Until we have probed the atmospheres of the giant planets, examined the surfaces and interiors of some of the moons and thoroughly explored Mars, we should not give up the possibility that life may have arisen more than once around our Sun.

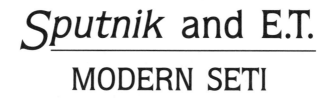

7

Sputnik and E.T.

MODERN SETI

Those who feel that the goal justifies the great amount of effort required will continue to carry on this research, sustained by the possibility that sometime in the future, perhaps a hundred years from now, or perhaps next week, the search will be successful.

Frank Drake (1961)

SPUTNIK NOT ONLY TRIGGERED the exploration of the solar system, it brought respectability to the search for extraterrestrial intelligence.

In 1959, just two years after *Sputnik*, a pair of scientists published an article in the reputable British scientific journal *Nature*, shaping the modern search for extraterrestrials. The scientists, Giuseppe Cocconi and Philip Morrison, investigated how another civilization might be detected by us, and concluded that the easiest way to communicate across the galaxy would be by radio signals. Imagine reading these words of Cocconi and Morrison not in a science-fiction magazine but in the pages of a journal more accustomed to the sober words of Nobel prizewinners:

Pioneer SETI theoretician Philip Morrison, at the 1985 turn-on of the META SETI system that embodied some of his ideas.

> It follows, then, that near some star rather like the Sun there are civilizations with scientific interests and with technical possibilities much greater than those now available to us.
>
> To the beings of such a society, our Sun must appear as a likely site for the evolution of a new society. It is highly probable that for a long time they will have been expecting the development of science near the Sun. We shall assume that long ago they established a channel of communication that would one day become known to us, and

that they look forward patiently to the answering signals from the Sun which would make known to them that a new society has entered the community of intelligence.

Several years earlier, such a paper probably could not have been published in a major scientific journal. But a transformation had occurred in the minds of many scientists, caused, I believe, by *Sputnik*. That tiny unmanned vehicle suddenly tore the concept of space travel out of the comic book and glued it into the textbook. Buck Rogers fantasy had become Walter Cronkite reality.

In this atmosphere, a few scientists began talking about subjects previously found primarily in journals such as *Astounding, Amazing, Galaxy* and *Fantasy and Science Fiction.* Cocconi and Morrison were two of these pioneers. They reasoned that rocket travel would be slow and expensive, at least for the foreseeable future. But they knew of one technology much faster and cheaper: radio.

Radio signals (and television, a special type of radio signal) travel at the speed of light, 186,000 miles per second. Even at that speed,

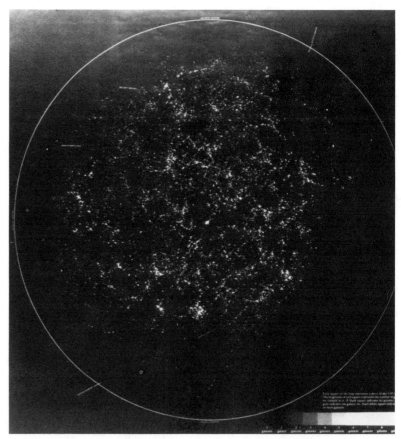

A computer image of one million galaxies. Each dot is a galaxy with, typically, hundreds of billions of stars. This is just a small sample of the galaxies in our cosmos, and the most fundamental reason for optimism in SETI. (The dark area to the right is due to a lack of data in this sample.)

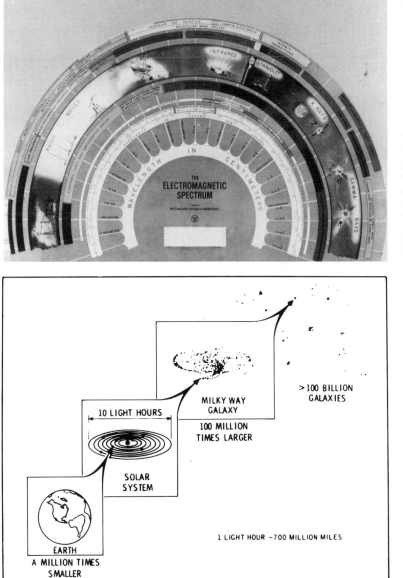

The electromagnetic spectrum. Waves of electromagnetism range from low frequency, long wavelength radio signals up to exceedingly high frequency, short wavelength light and gamma rays. They usually travel at the speed of light, and are the best way we know of to communicate over the vast distances between stars.

Our relationship to the universe.

it would take four years for a signal to reach us from our nearest stellar neighbor, Alpha Centauri, since it's four light-years away. Even in 1959, there were radio transmitters, receivers and antennas being built powerful enough to communicate across that distance.

Cocconi and Morrison then wondered which channel to tune to. On Earth, most radio and television signals are broadcast on well-defined channels. These signals are established by international agreements which, in the United States, are monitored by the Federal Communications Commission. Unfortunately, there is no Galactic Communications Commission, or at least none that we know of. This seemed to mean that the search would require aiming the antenna at each tiny piece of sky, then tuning through a practically infinite number of channels. It could take centuries to cover the whole sky.

The problem came to be known as the Cosmic Haystack, because it was similar to the proverbial search for a needle in a haystack. Cocconi and Morrison, however, pointed out that there's a natural interstellar channel. It stands out above all others in our galaxy. This channel is the fundamental radio signal hydrogen atoms broadcast.

Our galaxy is filled with hydrogen atoms. They're so thinly spread that they're about half an inch apart in interstellar space, a very good vacuum. But over the vastness of space, there are so many atoms that the feeble radio broadcast of each one adds up to a detectable signal. The atoms broadcast at a frequency of 1420 megahertz (1420 million cycles per second), less than twice the frequency of TV channel 83. If your TV set could tune to channel 172, and you hooked it up to a huge antenna, you would hear a hissing sound and see a lot of "snow"—Mother Nature's favorite TV show.

Cocconi and Morrison suggested that if there's a civilization out

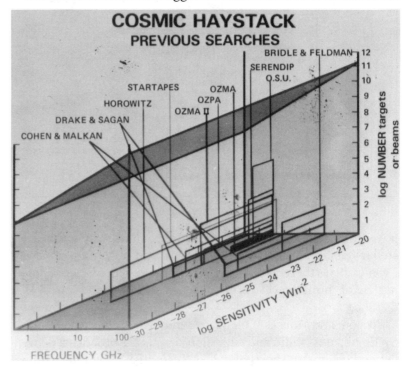

The cosmic haystack. SETI needs to search for an unknown frequency (horizontal axis) for a signal of unknown direction (vertical axis) and unknown strength (axis going into the paper). All SETI has examined so far are tiny slices of "search space," and could easily have missed a signal if it's there. (Frequency in gigahertz = 1000 megahertz; sensitivity in watts per square meter; direction is represented by the number of areas in the sky covered by the antenna used.)

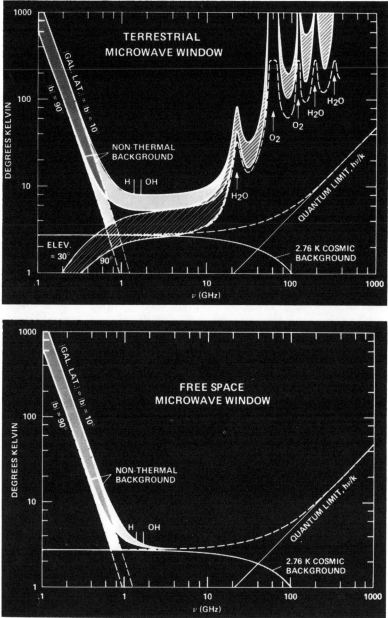

The spectrum of natural noise facing the earthbound SETI observer. The vertical scale is a measure of the noise produced by the radio emissions of the galaxy and of our atmosphere. The horizontal scale is the frequency in units of 1000 megahertz (1 gigahertz). Notice that the middle — the microwave region—is where it is quietest.

The noise spectrum of the universe, as it looks above our atmosphere. The vertical scale is a measure of the noise produced by the radio emissions of the galaxy and of our atmosphere. The horizontal scale is the frequency in units of 1000 megahertz (1 gigahertz). Again, the microwave region is where it is quietest.

there, and if they're at least as smart as we are, they already know about this hydrogen signal. If they want to make themselves known to other civilizations around the galaxy, the easiest way would be to broadcast at or near this frequency. It was nicknamed the "magic frequency," because it stands out so distinctly from the background noise, like a rabbit popping out of a conjurer's hat.

The two scientists concluded their article with this statement:

The reader may seek to consign these speculations wholly to the domain of science-fiction. We submit, rather, that the foregoing line of argument demonstrates that the presence of interstellar signals is entirely consistent with all we now know, and that if signals are present the means of detecting them is now at hand. Few will deny the profound importance, practical and philosophical, which the detection of interstellar communications would have. We therefore feel that a discriminating search for signals deserves a considerable effort. The probability of success is difficult to estimate; but if we never search, the chance of success is zero.

A Needle in the Cosmic Haystack?

One scientist soon learned what it felt like to be the first person to hear radio signals from an extraterrestrial civilization. Or so he thought.

Eventually he discovered that the signals were not from another civilization, but for two weeks he thought he had succeeded. His reaction was fascinating, and it gives us a glimpse of what it will be like when the first person successfully detects signals from another civilization.

The man was Frank Drake. Today, he is recognized as a SETI pioneer. But in 1960, he was just another young scientist working on an 85-foot-diameter dish antenna at the National Radio Astronomy Observatory in Green Bank, West Virginia. He was a radio astronomer, studying the radio signals produced by hydrogen and by other natural processes in our galaxy.

Independently of Cocconi and Morrison, he used similar reasoning. Drake conceived a project he called Ozma, "named for the queen of the imaginary land of Oz—a place very far away, difficult to reach, and populated by exotic beings." His historic experiment ushered in the present era of radio searches for signals from other worlds. This was the birth of modern SETI.

He pointed his antenna to two nearby Sunlike stars, Epsilon Eridani and Tau Ceti and listened for signs of artificial radio transmitters.

I spoke with Drake twenty-five years later at Caltech, where he was giving a lecture on SETI at a conference on the origin of life. Today, he is a silver-haired dean at the University of California at Santa Cruz. After a quarter century of SETI and radio astronomy, the memories of those early days are still sharp. "This was the first day of the experiment," he recalls. "We'd been running since six o'clock in the morning on another star, and heard nothing but noise. The first star was Tau Ceti."

At that time, it was normal for radio astronomers to use a chart recorder similar to a lie detector, with pens moving back and forth, writing in ink the radio noises of the cosmos. But in addition, they hooked up a loudspeaker to the radio so they could hear sounds with their own ears.

The big surprise came when they looked at their second star, Epsilon Eridani. "When Tau Ceti set," says Drake, "we moved to Epsilon Eridani, and as soon as we set up the receiver, within seconds, we heard this thing, which is a remarkable coincidence.

"What we heard were noise pulses. It didn't sound like a normal intelligent transmission, which has voice or tones or something. It was very strong pulses of noise, eight per second, actually. It persisted." Nothing remotely like that had ever been detected from the galaxy before.

What did he feel? "Great excitement. A sudden thought—'Was it really this easy?' "

It took Drake weeks to figure out what those pulses really were. He and his colleagues hadn't yet thought through how they would respond to such an incident, since they hadn't expected it to happen so soon. Their first reaction was to run over and check the equipment. So they spent about a minute checking things over, and everything seemed to be correct, at which point they realized that the right experiment was to move off the star and see what happened. So they did this and the signal went away.

The Howard E. Tatel 85-foot antenna in Green Bank, West Virginia, used by Frank Drake in Project Ozma.

When they went back to the star, the signal was not there anymore. So the question was, where had it come from? Had it turned off? For weeks they went back to that frequency and that star daily. For about two weeks, they were really hopeful.

They also realized that it could be interference. So they set up a second receiver with an omnidirectional antenna. If signals came in from any part of the sky, they would be detectable in that system as well.

A lesser scientist might have announced to the world the detection of extraterrestrial life, but Drake knew that anything so important had to be *extremely* carefully checked. Hence the two antennas: the 85-foot radiotelescope dish which can only receive signals from the point in the sky at which it's aimed; and an omnidirectional antenna of much lower sensitivity, able to pick up signals from anywhere in the sky. If the signal repeated, and came in on both the antenna aimed at the star and on the omnidirectional one, it would mean the signal was really just interference, not connected with the star at all.

About two weeks later, the signal did reappear. But it reappeared in both systems, which meant that it was just ordinary interference.

What was it?

"We never knew what it was," says Drake. "The second time it came on, we followed it and measured its intensity, and the way it built in strength, and receded. It appeared to be coming from an airplane that was flying over. People have told us, being concrete about it, that it was probably an airborne military jamming system that was being tested, and they'd gone to that remote part of West Virginia because they thought it would do the least harm. This is supposedly from people who really knew what it was, but it was classified so they couldn't really tell us."

Just as in *The Wizard of Oz*, the mysterious sorcerer turned out to be an ordinary guy with no special powers.

Frank Drake's experiences with Project Ozma typify the problems that have plagued SETI scientists ever since: the difficulty of distinguishing real signals from interference; the prevalence of vast amounts of radio pollution, some of it from people extremely reluctant to announce their activities to others; the pulse-pounding joy of apparent success; and the heartbreaking disappointment when the needle in the cosmic haystack turns out to be nothing but a thorn.

The Drake Equation

After conducting the first modern scientific SETI program, Frank Drake also developed a way of estimating the number of extraterrestrial civilizations. How can you possibly calculate something so filled with unknowns? Eventually, after a lot of discussion with his colleagues, he broke the question down into seven steps. Many of the numbers used in these steps are hotly debated, so I'll use rough figures that I think are reasonable.

Step 1. How many stars are there in our Milky Way galaxy?
Easy. Astronomers estimate there are about 400 billion stars here.

Step 2. How many of these stars have planets?
Tougher. But we have seen that there is good reason to believe planets are common, so let's estimate that perhaps one star in every ten has planets. That would mean there are around 40 billion stars with planets.

Step 3. How many of these planets are suitable for life?
First, we need to have some idea of how many planets per star there are. All we know for sure is that there are at least nine planets around our own star. The prevalence of dozens of moons around Jupiter, Saturn and Uranus, each one resembling a miniature solar system, suggests that multiple planets might be common, too. The theory of planet formation also suggests that if planets can form

Frank Drake and colleagues have occasionally used the Arecibo radiotelescope for SETI. A view from the bottom of the dish, looking up through the triangular frame from which radio antennas are suspended.

at all around a star, there will be enough material spread around to allow many planets to form, not just one large, economy-size world. So let's estimate that whenever planets form around a star, roughly ten planets per star are created. Thus there should be something like 400 billion planets.

But how many of these are suited to life? If we are conservative and limit our thinking to life as we know it, then that means we need a planet that is not too hot and not too cold, with a moderate atmosphere and substantial water. In our solar system, we know of only one such planet for sure, though Mars and Venus are very nearly suitable, and had their histories been slightly different, either of them might now be graced with oceans and life. And there are the other potential niches on the various planets and moons mentioned earlier. But let's be conservative and estimate that only one planet per stellar system is usable. That makes 40 billion Earthlike planets.

Step 4. How many of the nice planets actually develop life?

That's tough. Some scientists think that if you just let a watery planet sit there at moderate temperatures for a few billion years, life will automatically evolve. Others are much more skeptical and think that life is very difficult to start. Let's suppose, for the sake of argument, that one such planet in ten will develop life. That still gives us four billion planets with life. If true, the galaxy is crawling with life!

Step 5. How many of those develop intelligent life?

That, of course, is even more speculative. I personally suspect that once life has started, it will evolve more complex creatures of such an incredible diversity that, sooner or later, one of them will develop intelligence. If we had not arisen, many other creatures might have developed brains and grasping organs powerful enough to manipulate the environment: dogs, cats, bears, opossums, raccoons, otters, elephants, dolphins, whales, octopi, dinosaurs. Once even a primitive intelligence arises, provided that it has arms or fingers or tentacles with which to manipulate the environment, it has a tremendous advantage over its competitors. From that day onward, competition within its own species will cause the smartest ones to win out over the duller ones and true intelligence should be the end result. But some think that intelligence may not evolve often. So let's say that one in a hundred planets with life evolves intelligence. That's 40 *million* civilizations. We could have a lot of neighbors.

Step 6. How many of those develop civilizations with technology capable of interstellar communication? (There could be whole planets full of intelligent manicurists, but if they don't have radio, we cannot detect them with radio SETI.)

Some civilizations might develop in very different directions from

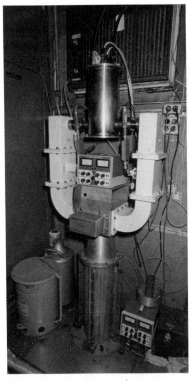

One of several gallium arsenide field-effect transistor receivers used at Arecibo.

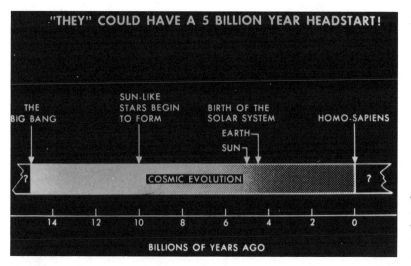

They may be far ahead of us. Our planet has only been around for five billion years; they could have reached our level of technology billions of years ago.

ours. They might concentrate on arts, crafts, philosophy, biology. They might never be able to make a simple radio transmitter. But since astronomy has shown that the laws of physics are the same everywhere in the universe, and since curiosity seems to be a natural byproduct of intelligence, it is reasonable to assume that many creatures, once they have become intelligent, will discover the same physical laws we have, and eventually build devices able to communicate over interstellar distances. If even one civilization in ten does this, there might be four million technological civilizations.

Step 7. But how long do those civilizations last?

Pessimists think our own civilization may not last much longer. If they are right, then we will have been able to communicate with other civilizations for only a century or so. It is ironic that at virtually the same moment in history when technology reaches the point that we can communicate with other worlds, it also enables us to blow our own world up. But if the optimists are right, we will be around forever. Even then, we may eventually get bored with technology and become uncommunicative once again.

Some civilizations may become very introverted. Suppose a way was found to stimulate the pleasure centers of the brain through narcotics or electricity, and that life-support equipment allowed beings to remain hooked up in perpetual ecstasy, while robots took care of maintenance. There could be whole planets full of happy, uncommunicative pleasure-addicts. Still, the diversity of our species suggests that not everyone would choose such a meaningless existence, and if just a few beings refused to vegetate, with this level of technology, they could easily communicate across the interstellar gulf.

I suspect there may be civilizations that die—or cease being technological—but their machines live on. Advances in artificial

The Earth being covered with the soot of World War III, triggering Nuclear Winter. If this is the fate of many civilizations, then even if intelligent life arises frequently, there may not be many who live to tell the tale.

intelligence seem to occur every day, making our computers smarter and smarter. The era of the self-repairing machine has already dawned—spacecraft such as *Voyager* automatically switch in backup units when the main circuits fail. Eventually, self-repairing machines may be common, as well as robots to fix those that cannot repair themselves.

One day, if we destroy ourselves, our machines might live on, becoming our mechanical descendants. There could be whole worlds of robots out there, patiently signaling, the final embodiment of a culture dead a billion years. Civilizations may achieve immortality through their machines.

Let's look at some possible numbers. The Earth is 5 billion years old and the age of the universe is about 15 billion years, so the average age of a planet is probably around 10 billion years. Suppose a civilization is communicative for a thousandth of the age of its home world—10 million years. That would mean that one-thousandth of the 4 million technological civilizations that arose—some 4,000 worlds—could be out there *right now*, waiting for us to detect them.

Drake put all of these concepts into one equation for N, the number of advanced technical civilizations in the galaxy:

$$N = N_* \, f_p \, n_e \, f_l \, f_i \, f_c \, f_L$$

where N_* is the number of stars in the Milky Way, f_p is the fraction of stars with planets, n_e is the number of planets ecologically suitable for life, f_l is the fraction of those planets where life evolves, f_i is the fraction that develops intelligent life, f_c is the fraction that communicates, and f_L is the fraction of a planet's life occupied by the communicating civilization.

Thus Drake, in one sweeping equation, encompassed everything from our evolution to our possible extinction. Of course, pessimists put smaller numbers in each of those steps and get much smaller numbers of civilizations, even less than one. If they are right, then we may be alone in this galaxy—but there are billions of other galaxies in the universe, so not all is lost.

I side with the optimists, who put larger numbers in each step. Many of the discoveries of the last decade have supported the optimists: the discovery of rings around stars; the increasing complexity of molecules detected in interstellar clouds; the finding of warm spots in the cold reaches of the solar system; and the discovery of tubeworms and other exotic lifeforms radically different from all other Earth-life, near hot, deep volcanic vents in our oceans. If we optimists are right, there are thousands or even millions of civilizations communicating right now, and all we have to do is be clever enough to detect them.

The Drake equation shows clearly one of the important consequences of SETI: it forces us to think about our past and future from a cosmic perspective.

Robert Dixon at the Ohio State University radiotelescope, the site of the world's longest-running SETI program.

E.T. vs. the Golf Course

Project Ozma and the Drake Equation inspired other SETI searches, and SETI encountered unexpected obstacles and produced several unsung heroes. Two of the heroes who encountered such an obstacle are astronomers John Kraus and Robert Dixon of Ohio State University.

For a decade they've been conducting the world's longest continuously-run SETI program with practically no money, using old equipment, unable for a long time even to afford a modern minicomputer. They also started one of the world's most unusual journals, the first magazine devoted to SETI, *Cosmic Search* (which has since, unfortunately, folded). In recent years, they have received some support from NASA and The Planetary Society, but to this day they still struggle constantly to find financial aid.

As if the vastness of the Cosmic Haystack and the difficulty of research funding were not enough, they woke up one morning to face an even worse problem, stemming from the fact that a different university owned the land their radiotelescope is on. That university sold it without warning—to a real-estate developer who wanted

SETI scientists Robert Dixon and Michael Papagiannis with the Earth Flag.

A chart of the universe as seen by the Ohio State University radiotelescope.

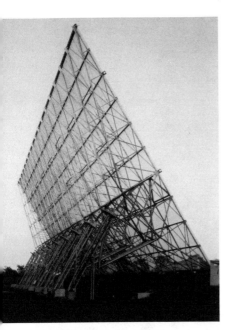

The Ohio State University radiotelescope, the world's longest-running SETI program.

to expand the golf course adjoining the research site. This would have obliterated the telescope.

The battle between the search for extraterrestrials and the developer was irresistible to the press, and soon the story was carried internationally. Officials in Ohio, hoping to attract high-tech industry to the state, were embarrassed. Finally, after much horsetrading, an agreement was reached that permits the observatory to continue functioning in return for an annual payment.

Kraus and Dixon thereby earned the dubious distinction of running the only SETI observatory in the world that has to pay a golf fee.

Implications of Ozma

Project Ozma crystallized the thinking about our place in the universe. It became inescapable that alien civilizations may be signalling right now. Still others may have died out in an orgy of self-destruction.

Just thinking about the problem of detecting other civilizations has already given us profound insights into our origin and destiny. SETI leads us down many unexpected paths: the history of the Earth, the motives of alien civilizations, the possibilities of advanced technologies, the destruction of worlds.

SETI brings together some of the greatest questions philosophers have ever asked. And—unlike philosophy—it offers the possibility that some of these questions may be answered in our lifetime.

Interstellar Postcards

MESSAGES TO SPACE

One picture is worth ten thousand words.
Chinese proverb

FOR CENTURIES, a lonely sailor or an adventuresome landlubber put messages into bottles and tossed them into the ocean, hoping someone might find them, thousands of miles away. This is just what we have done, on a much larger scale, with radio signals and with spacecraft. A message to other civilizations has been broadcast, and four spacecraft have been launched that are now leaving the solar system.

How do you write a message for someone who knows no human language? In the 1960s, Frank Drake devised a clever approach to the problem. He wrote a short message containing some basic information we would like to know about another civilization, putting it as a series of ones and zeroes of the sort used by computers. Then he mailed copies to his colleagues, asking them to decipher it. They had great difficulty, but the mathematically inclined reader is encouraged to attempt to make sense of this message.

If you give up, here's the answer: the message is a television picture. To see the picture, you have to notice that it contains 551 characters. This number is special, the product of 19 and 29. In turn 19 and 29, are even more special numbers, called *prime numbers*, numbers that cannot be evenly divided by any other numbers except one or themselves. Thus 551 equals 19 times 29, and cannot be written any other way.

This means that a rectangular picture can be built up from the 551 digits by putting them down in 29 rows of 19 characters, put-

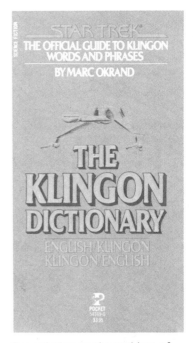

One solution to the problem of conversing with an alien civilization: a dictionary. Without such a book, we will have to rely on pictures and the universal laws of physics and mathematics.

119

Frank Drake's hypothetical alien
message. Can you decipher it?

```
1111000010100100001100100000000100000010100
1000001100101100111100000110000110100000
0010000010000100001000101010000100000000
0000000000100010000000000101100000000000
0000000100011101101010110101000000000000000
0000100100001110101010101000000000101010101
0000000001110101010101110101100000001000000
0000000000100000000000001000100111111000
001110100000101100000011100000001000000000
1000000001000000011111000000101100010110
100000000110010111110101111100010011111001
0000000000011111000000101100011111111100000
100000110000110000100001100000000011000101
001000111100101111
```

ting black squares where there are ones and white squares where
there are zeroes. The picture shows a crude image of a humanoid,
together with information about chemistry, astronomy and biology.

Admittedly, it's hard to understand his message, but a committee of scientists could figure it out. In fact, he found that individual
scientists usually were able to understand only the parts of the
message related to their own specialties. If we were not limited to
short messages, we could send more detailed images and avoid the
uncertainties of this crude picture, and in fact, since then, increasingly sophisticated messages have been sent into space.

Our first deliberate note in a space-bottle was on NASA's *Pioneer
10* spacecraft. It was designed to be the first vessel to fly by the planet

The meaning of Frank Drake's
hypothetical alien message: it's a
picture. There's a figure of a
humanoid; along the left border is
a picture of its star and nine
planets. The upper right corner
shows diagrams of carbon and oxygen, suggesting that they have a
similar chemistry. Next to the first
five planets are the first five whole
numbers, written in the binary
computer style, with an extra bit
added for error detection: 10 = 1,
100 = 2, 111 = 3, 1000 = 4, 1011
= 5. (The extra digit, called a pari-
ty bit, causes each number to have
an odd number of ones.) There's a
cartoonist's "balloon" connected to
the alien by the diagonal line.
Evidently, it is telling us the three
numbers in the balloon. The top
number, in line with planet 2, is
11; the next, near planet 3, is

about 3,000, and the last, near
planet 4, is about 7 billion. It says
that they have 11 beings on planet
2—evidently an expedition; 3,000
on planet 3—a space colony; and 7
billion on planet 4—their home. To
the right of the creature, its height
is marked off by the number 31. It
is 31 units high. The only unit of
length measurement in the
message is the wavelength of the
radio signal. The four-square symbol below the creature is the alien's
code for themselves, for future
messages. (It can't be a number,
because it doesn't have an odd
number of ones.)

If you had trouble, don't blame
yourself. It would take a team of
scientists with varying specialties
to completely understand such a
message.

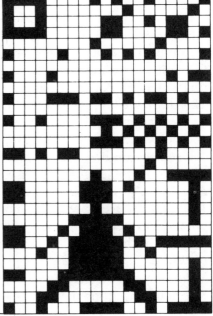

Jupiter, but it had a distinction that made it one of the most extraordinary events in human history: *Pioneer 10* was the first object from our civilization that would leave the solar system forever.

In giving the rocket enough of a boost to get to Jupiter, we left it with enough energy to continue sailing forever. It would not only have escape velocity from the Earth, it would have escape velocity from the Sun, after Jupiter's gravitational pull.

On realizing that *Pioneer* would become the first object in the history of the human species to leave our planetary system, space writer Eric Burgess came up with the idea that we send a message to any alien civilization that might find the spacecraft, even millions or billions of years in the future.

Burgess and freelance writer Richard Hoagland contacted Carl Sagan, who greeted the idea enthusiastically. Sagan and Frank Drake designed a plaque that any advanced being should be able to decipher. Sagan's wife at the time, Linda Salzman Sagan, drew the human figures for the plaque. NASA approved the idea and etched it onto a gold-anodized aluminum plaque mounted on a part of the spacecraft that is shielded from interstellar dust.

Naked figures of a man and woman on the plaque show the creatures who built the probe, prompting some angry mail contending that NASA was sending out space pornography. It is doubtful,

"Take me to your stove? . . . You idiot! Give me that book!"

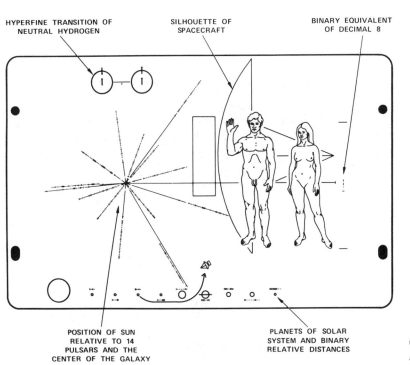

HYPERFINE TRANSITION OF NEUTRAL HYDROGEN

SILHOUETTE OF SPACECRAFT

BINARY EQUIVALENT OF DECIMAL 8

POSITION OF SUN RELATIVE TO 14 PULSARS AND THE CENTER OF THE GALAXY

PLANETS OF SOLAR SYSTEM AND BINARY RELATIVE DISTANCES

Our first interstellar postcard: the *Pioneer* plaque.

An artist's view of the *Pioneer* spacecraft after it has wandered millions of miles through space, battered by micrometeoroids but still bearing its message.

however, that the two-headed space-faring octopi or others of that ilk who come upon it will find the picture noticeably erotic.

Pioneer 10 and its twin, *Pioneer 11*, with the same plaque, were successfully launched in 1972, starting their interstellar postcards on the longest journey in history.

Interstellar Telegram

In 1974, scientists from Cornell University broadcast the first deliberate message to space. It contained, in a mathematical code, a crude picture of us and basic information about our chemistry. They broadcast it from Cornell's giant radiotelescope antenna in Arecibo, Puerto Rico, taking advantage of the powerful transmitter it has in addition to its receivers. Normally, the transmitter is used for radar, bouncing signals off the Earth's ionosphere or off other planets. But on this occasion, it was used to beam the message to the Great Cluster, 300,000 stars in the constellation of Hercules, 25,000 light years away.

Carl Sagan describes the message this way: "What it said fundamentally was: Here's the Sun. The Sun has planets. This is the third planet. *We* come from the third planet. Who are we? Here is a stick diagram of what we look like, how tall we are, and something about what we're made of. There's four point something billion of us, and this message is sent to you courtesy of the Arecibo telescope, 305 meters in diameter."

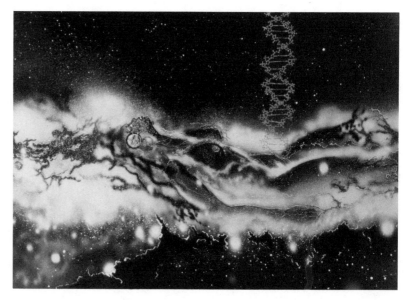

Space artist Jon Lomberg's fanciful image of the DNA molecule transmitted to space, witnessed by the Woman of the Milky Way.

Twenty-five thousand years from now, if there is anybody home there, they may reply, and it will take another 25,000 years for their answer to come back. In the game of two-way interstellar communication, patience is a much-needed virtue.

Interstellar Jazz

In 1977, the next probes to Jupiter would also leave the solar system. These were NASA's *Voyager 1* and *2* spacecraft. To top the *Pioneers'* act, they carried a kind of videodisk, designed by Sagan, Drake and their colleagues. Encoded in the grooves of this disk are some of the sights and sounds of Earth, along with lots of scientific information.

It contains recordings of greetings from the Secretary-General of the United Nations, Kurt Waldheim, and President Jimmy Carter. Messages in dozens of languages are given, as well as photographs of people and places from all over the planet and music from Bach to Chuck Berry, including representative music of countries from all over the world. It's been called "Earth's Greatest Hits."

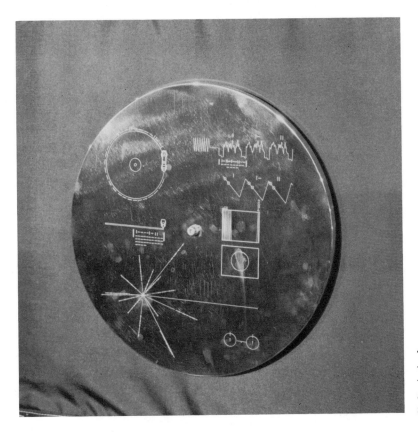

The *Voyager* disk. This is actually the cover of the disk, with instructions on how to play it. Inside, the actual disk contains pictures and sounds of Earth.

Interstellar Sitcoms

These messages are not the only ones we've sent to space. The Arecibo message is the only modern, scientific message we have deliberately broadcast, but there are two other types of signals we have been unintentionally broadcasting for decades. The most powerful interstellar signals our civilization routinely sends are our TV shows and our military radar.

I Love Lucy, Hee Haw, Bowling for Dollars and the six o'clock news are now all traveling into space. These pinnacles of human culture are there for any nearby civilization to see and to attempt to decipher.

Shows broadcast four years ago are just now reaching Alpha Centauri, since it is four light-years away. Lucky Alpha Centaurians are being treated to *Love Boat, Dallas, Laverne and Shirley* and, appropriately, *That's Incredible.*

Shows broadcast forty years ago are now forty light-years away, a volume of space including millions of stars. Those at this frontier of human civilization are now enjoying the *Gillette Cavalcade of Sports, Cash and Carry* (a quiz show set in a grocery store) and for intellectual sustenance, *I Love to Eat*, a cooking show hosted by James Beard. Let's hope that none of the things that the host of the latter loved to eat resemble too closely any of his viewers forty light-years from here.

"It's a message from the spiral nebula in Andromeda. They want us to play Bruce Springsteen!"

Many scientists who have pondered these broadcasts have suggested that aliens who study our signals will conclude there is no intelligent life on Earth.

Interstellar Insights

Why send messages into space when we know we are not likely to receive an answer in our lifetime? Partly to answer the critics who say that interstellar distances are so great, and the time for messages to travel so long, that no civilization would ever do it.

The *Pioneer*, Arecibo and *Voyager* messages prove that we do it. And if we do it, using primitive twentieth-century technology, then other civilizations may do it with more powerful techniques.

Of course, it is potentially more rewarding to search for signs of other creatures. Harvard SETI researcher Paul Horowitz says that one of the main reasons for receiving is that "you're impatient to receive, because if you send, you know darn well you're going to have to wait a round-trip travel time, at least ten or twenty years. And experiments that take ten or twenty years are intrinsically less interesting than ones that might return an answer tomorrow."

But it must be pointed out that ten or twenty years is a very optimistic number. As Carl Sagan observes, even the most optimistic estimates of the distance to the nearest civilization generally come out to be hundreds of light-years away. "Which would mean," says Sagan, "that if you were saying 'Hello' to a lot of guys, it would be, you know, 2511 or something when the message would get back. It might be a little disappointing for you personally to take that long."

Few people realize it, but our first interstellar spaceship is now on its way to the stars. On June 13, 1983, the tiny *Pioneer 10* spacecraft crossed the orbit of Neptune and left the solar system. (Normally, Pluto is the farthest planet from the Sun. However, for about twenty years out of every 248, Pluto's highly eccentric orbit takes it just inside Neptune's. Thus, from 1979 to 1999, Neptune, not Pluto, is the most distant known planet in our system.)

It took us millions of years to get from foot travel to heavier-than-air flight. It took half a century from the Wright Brothers to *Sputnik*. In just a quarter of a century, humanity went from Earth-orbiter to interstellar travel.

Our lonely civilization, living on this island called Earth, has now sent out four interstellar spacecraft and countless radio and TV signals, our interstellar bottles, adrift on the galactic sea. Perhaps one day, someone or some *thing* will find one of these messages.

"As I understand it, they want an immediate answer. Only trouble is the message was sent out 3 million years ago."

The structure of the DNA molecule transmitted from Earth.

They will puzzle over it, apply logic, make some guesses. They may enjoy the crude craftsmanship of our primitive spacecraft much as we admire native basket weaving. They may find the antics of *I Love Lucy* and the six o'clock news equally comic.

And just possibly, they may send back the equivalent of "Having a wonderful time. Wish you were here."

Little Green Men and Nobel Prizes

FALSE ALARMS

Probably he who never made a mistake never made a discovery.
Samuel Smiles

THE HISTORY OF SETI is studded with false alarms, moments when alien signals seemed to have been detected but were proven to be wrong. Marconi's apparent detection of natural whistler waves and Frank Drake's first Ozma signal were two of the earliest.

In 1967, an event occurred that was the closest thing yet to a detection of another civilization. The events of those days show us the excitement and the pitfalls of the search, and give us a preview of what may happen if SETI really is successful.

A young Irish colleen from Belfast by the name of Jocelyn Bell was at the controls of a new Cambridge University radiotelescope in England. Her father had been the architect of the Armagh Observatory in Northern Ireland, and the astronomers there encouraged her adolescent interest in their field.

She entered the Ph.D. program at Cambridge University and helped complete a radioastronomical antenna resembling hundreds of dissected TV antennas spread over five acres of ground. The array was hooked up in a way allowing it to scan a large chunk of Her Majesty's sky, and was intended to study the familiar run-of-the-galaxy natural radio sources that populate the universe.

Bell was a member of a team of researchers under the direction of Dr. Antony Hewish, and she noticed some odd pulsating noise being picked up by the receiver, like the kind that spoil TV programs while turning radio astronomers' hair gray. It wasn't the sort of thing you'd seriously expect to be anything more profound than in-

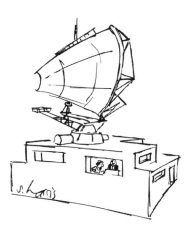

"As I read it, we're receiving a message from outer space telling us to stop bombarding them with unintelligible messages."

127

terference from someone's Model T, but they soon noticed that the signals emanated from a point in the sky fixed with respect to the stars. This ruled out the possibility of an Earthbound radio source, but with all the hardware tossed into the skies by space programs, who was to say it wasn't one of our space probes?

There was a good precedent for their caution. They were well aware of Frank Drake's Project Ozma, and of how easily false alarms come to those looking for signals from another civilization.

History didn't repeat itself. The source was not human. Hewish and his colleagues observed the signals for many months, and found that during this time the position of the radio source remained unchanged in relation to the stars. This ruled out the possibility that the source was a manmade space probe, radar signals bouncing off the Moon, or any other object in our solar system.

They naturally wondered if any more such oddball sources existed. Knowing now what kind of signals to look for, they quickly found three more *pulsars*—as they later called these objects—in different parts of the sky, with remarkably similar behavior.

The Hewish group held off the announcement of the discovery until their paper appeared in the February 24, 1968, issue of the British scientific journal, *Nature*, under the title, "Observation of a Rapidly Pulsating Radio Source." Then, pandemonium reigned in the astronomical world. (This may be an exteme way of describing the scientists' emotions, but when a scientist says "Rather interesting, that," it's equivalent to anyone else whooping jubilantly and jumping for joy.)

For some time thereafter, astrophysicists ignored their *Natures* and their *Astrophysical Journals* in favor of a glimpse at the latest hastily-Xeroxed *New York Times* and *London Times* interviews with the pulsar people. They then sat around trying to decipher the layman's language and the typographical errors to figure out what the interviewed scientists had really meant.

The discoverers did something that irked many radio astronomers around the world. Their original paper described the existence of four pulsars, but they neglected to mention the positions of three of them. To an astronomer, this is like saying that somewhere in the Sahara there are four bags of diamonds for the taking, and giving the location of only one. One bag is infinitely better than none, but four would be so much nicer—especially if they provided clues to the location of a whole diamond mine. So the rest of the scientific community was kept chewing its fingernails and had to be content with the table scraps of the one pulsar the Hewish people had already analyzed to death, while the Britons observed the secret three at their leisure.

"Other laboratories tried to bargain for the valuable coordinates by offering privileged information of their own," reported Harold L. Davis, the senior editor of *Scientific Research* magazine. "Security at one center was so tight that theorists working there could only learn about the results of experimentalists at their own laboratory by communicating with scientists on another continent." Eventually, the Hewish group released the positions of all their pulsars, allowing everyone to get down to the business of serious pulsar watching.

Competition

The world's best pulsar receiver was the one run by Cornell, at Arecibo. I was a graduate student of astronomy there at the time, and I remember vividly the excitement of astronomers at the discovery of pulsars.

Frank Drake, the Project Ozma man, was in charge of Arecibo, which has the world's largest radiotelescope dish, making it the most sensitive receiver in the band for which it's designed—which just happens to be where pulsars are strongest. The dish focuses the signal collected by its huge area onto a jungle of TV-like antennas at its focal point several hundred feet above the ground. Many radiotelescopes steer as a single unit, like scaled-up versions of ships' navigational radars. To move the Arecibo dish, you'd have to move a large chunk of Puerto Rico, so instead they designed the dish in a way that allows them to aim the telescope by moving the much smaller focal antennas around.

Describing the big dish to an audience of scientists at M.I.T.'s Lincoln Laboratories, Drake once said, "The effective collecting area is about equal, surprisingly, to the combined collecting area of all the microwave and optical telescopes ever built in the history of man. The bowl would hold 357,000,000 boxes of corn flakes—or four billion bottles of beer!"

It was thus something of an embarrassment to Cornellians that their pride and joy, the Arecibo telescope, had not been the one to discover the pulsars. But it's a general property of antennas that the larger the dish, the narrower the part of the sky that can be seen, so the longer it will take you to search the sky. Thus, even though Arecibo is a superb instrument for viewing known pulsars, it's an incongruously poor one for discovering them. So narrow is its view that astronomers there were at first unable to detect the original pulsar even though they had Hewish's data. The slight inaccuracy in the pulsar position, combined with the erratic behavior of the source and the lack of knowledge of the best frequency at which

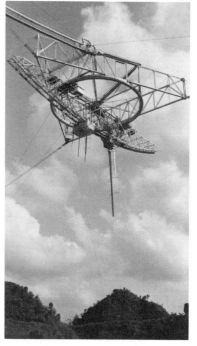

The antennas where the signals are focused at Arecibo. The curved "rocker" is 300 feet long; the suspended structure, 630 tons. To work on the equipment, intrepid scientists, technicians and graduate students climb out onto this apparatus, 450 feet above the dish.

Small antennas see larger areas of the sky than big ones, but with much less sensitivity. Radio waves, sound waves, and light waves behave similarly—small antennas or speakers or optical telescopes can't focus as well as large ones. This is one of the reasons why a small portable stereo can't give as good a separation of musical instruments as a large home unit, and an amateur astronomer's telescope can't see details as fine as those seen by Mount Palomar. Also, a big mirror gathers more energy, so can detect fainter signals.

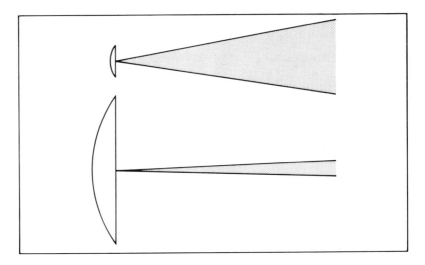

to tune their receiver, all conspired to frustrate their first efforts to detect the signals. They easily detected the pulsar a couple of days later when they tried a different frequency.

Drake tells of the hasty purchase, for the focal point of the multi-million dollar dish, of an antenna perfect for receiving pulsar signals—one on which untold kilobucks of research had been spent, although for other purposes than pulsar-reception. It was a humble, off-the-shelf TV antenna from a local hardware store!

Listening to pulsar signals is a strangely moving experience. The slowest pulsars sound just like the beating of a human heart, and you can't help at first wondering if these are the sounds of an inconceivably advanced alien civilization, received on Earth centuries after their broadcast. The fastest pulsar is much livelier, sounding like the beat of African tribal music. As Drake said, "You can dance to that one."

Green Beings?

The public wasn't ignorant of the excitement. At first the world was presented with a rash of press reports from scientists announcing that this looked like the real thing at last: the long-awaited signals from little green men. Privately, scientists even dubbed them "LGM objects," in honor of the little green men.

There was a lot of serious debate among scientists about the possibility that these pulsars might be signals to us from alien civilizations. Several arguments were made against that theory. The reasoning used to analyze the question is highly instructive, because

SETI will have to use the same type of logic if it ever again seems to have detected extraterrestrials.

One of the first anti-LGM arguments put forward was that the signals were definitely not being emitted from a planet. We can tell this is so because a planet, during its orbit around its parent star, would be moving toward the Earth during half its orbit, and away during the other half. Such motion gives rise to a Doppler effect, the same phenomenon that reddens the receding galaxies.

The pulses would become slightly more frequent if the planet were approaching, and less frequent if receding. So sensitive is this effect that the Hewish group found they could detect the Doppler effect of the Earth's orbital motion as we approached or receded from the pulsar. No such effect was found that could be attributed to a pulsar-planet, but this didn't rule out the LGM theory. It simply meant that, if the LGM theory were correct, the little green broadcasters had engineered their transmissions to artificially cancel out any Doppler effect that might annoy their listeners—or they simply might not have been broadcasting from an orbiting planet.

Astronomers could detect the Doppler effect due to the Earth's motion, despite the minute size of the effect, simply because of the remarkable constancy of the pulses. The ticks from the first pulsar occur precisely every 1.33730109 seconds, accurate enough to make a clock that gains or loses less than one second in a year. Two of the pulsars tick about as fast, but the fourth ticks about four times a second. They all tick with the same sort of fantastic accuracy, an accuracy that rivals that of atomic clocks.

Unlike the cases for flying saucers, the pulsar puzzle presents us with lots of concrete, objective, and above all, repeatable data. The problem lies with the interpretation of the data, and for this you must put yourself in the position of the astronomer.

The ideal astronomer has at least two qualities, inquisitiveness and practicality. His credo is: "If it radiates, measure it; if it doesn't, ask it for money." Pulsars are ripe candidates for measurement, so the first thing he wants to know about them—apart from the question of whether they're the residence of extraterrestrials—is how far away they are. Until you know the Sun's distance from us, you don't know whether it's a big, bright object zillions of miles away, or a small, faint one nearby. So, too, we need to know the pulsar's distance in order to learn whether it's emitting weak signals easily within the capacity of a reasonably advanced civilization's reception, or emitting huge quantities of energy inconceivable for any plausible civilization.

Hewish's people figured out the pulsar distance quite cleverly. They knew it was outside the solar system when they found that the source remained fixed with respect to the stars. An object in our solar system would seem to wander among the stars during the course of a year because our viewing platform, the Earth, moves around the Sun at the same time, changing the point of view from which we see the heavens. Stars are generally exempt from measurable changes because of their vast distances, just as your view of a nearby house against distant mountains changes greatly as you move past it, while your view of a house halfway to the mountains changes very little. This is the phenomenon called *parallax*.

Now they knew where the pulsars were not, but they wanted to know where they were. That, more or less, is what the Hewish people asked themselves and they came up with a very subtle answer. They noticed that a pulse observed on one frequency was delayed on another. The wider the difference in frequencies, the greater the delay. If you hooked your TV up to a huge antenna aimed at a pulsar, you'd get one pulsar's pulses on channel 2 about twenty seconds after those on the higher frequency of channel 13. How can this possibly allow you to figure out the pulsar distance?

Elementary, my dear Watson—elementary plasma physics, that is. It happens that interstellar space is swarming with electrons, and electrons slow down the radio waves. Space is a plasma—a gas in which the atoms are broken down into charged particles (electrons and protons, mainly). Interstellar gas typically has several electrons for every matchboxful of space. That may not seem like many, inasmuch as you've got a trillion trillion of them crowded in your little finger, but there are enough electrons in the Milky Way to account quite nicely for the time delay of the pulses.

The electrons, being very light and electrically charged, are easily wiggled by the electromagnetic waves of radio signals. High-frequency signals have only a slight effect on the electrons, but low frequencies push the electrons around strongly and in turn are delayed by the electrons they push. The farther away the pulsar, the more electrons the radio signal encounters and the longer the delay. Thus the low-frequency time-delay in the signals tells us how far away the pulsars are.

Plot the observed time delay versus frequency on a graph, and you get a straight line exactly as plasma theory predicts. In fact, so straight is that line that it's shocking to an astrophysicist. Astrophysicists are accustomed to obtaining graphs in which the errors are huge and the points scattered almost randomly over the paper. One astrophysicist will stare at such a graph, in which the data points are distributed as linearly as raisins in a bowl of cereal, and he'll say, "Those points obviously occur around a straight line with a slope of minus 1.4." A second astrophysicist will stare at it intently, and say, "No, I'm afraid it's minus 1.5." Another will point out that it looks like two intersecting straight lines, and the next will assure us that it's a curve.

Thus you can imagine the gasps of awe which greeted the display of this indisputably straight pulsar line, precisely as predicted by theory. Curves like that just don't happen in astrophysics. It's almost indecent.

Having measured the time delay, they were able to compute that the first pulsar was several hundred light-years away. If it had been any farther, the time delay would have been greater, and if nearer, shorter. The other pulsars were at similar distances.

Objections to Little Green Men

Knowing how far away they were, we could figure out how "loud" they were. Their signals are so strong on Earth (stronger at some frequencies than most other natural radio emitters in the sky, although much weaker than local TV and radio signals) that they must radiate about a billion times as much energy as is generated by the entire electrical production of Earth—assuming the signals aren't beamed directly at us.

This brings us to the first major objection presented against the intelligent-alien theory of pulsars—namely, that those little green men couldn't conceivably generate such huge amounts of energy. This is strongly reminiscent of those who said you couldn't possibly build an engine to pull a heavier-than-air craft through the air, or that it was impossible to build a rocket powerful enough to escape the Earth's gravity.

Present electrical production exceeds that of eighteenth-century Earth by a rather large percentage, and it is difficult to foresee by how much our electrical generation will have increased after another couple of centuries, let alone the billions of years the LGM's may have on us. Electrical production certainly isn't decreasing.

Of course, a billion earthpower is a lot to ask of even the most industrious extraterrestrials, but there happen to be two convenient ways around this obstacle. The first is to suppose that the pulsar signals are beamed at us. Presumably the members of an advanced civilization have a similarly advanced astronomy and biology, giving them a good idea of which stars are the most probable abodes of life. They might then choose likely candidates and beam their signals at them with a hundred-mile-diameter super-Arecibo dish. With such a superdish, they would need only about one earthpower to produce the signals we observe. A hundred-mile mirror sounds a bit tricky, but it need only be a gossamer wire mesh floating in space.

But suppose the LGM's don't know—or more plausibly, if ego-deflating, don't care—about us. It's a bit presumptuous of us to assume the pulsar inhabitants care about the aboriginal savages on some distant, uncivilized island in the galaxy, so is there any conceivable source of energy that could supply the billion earthpower needed if the signals were unbeamed? You bet your life there is. In fact, you bet your life every day on the continued existence of this powerhouse: the Sun. The Sun produces a hundred *trillion* Earthpower, a hundred thousand times the needs of those power-hungry little green men. How do you convert this energy, which is mostly in the form of light waves, into radio waves? That's their problem.

We don't have to pass the buck completely, because astrophysicists have already found that nature has set up huge *masers* (lasers using microwaves instead of light waves). Some interstellar clouds are natural masers, that generate strong radio waves. The theory of natural masers has already been advanced to explain another bunch of mysterious radio signals. These signals emanate from *OH regions*, regions containing water vapor in which each molecule lacks one of its hydrogen atoms: OH, called the hydroxyl molecule. If astrophysicists can conceive of radio-broadcasting maser steam-engines occurring as a mere accident of nature, then surely we can be granted the latitude of supposing that an ultra-civilized world could construct a similar maser deliberately.

It's amusing to recall that masers and lasers didn't exist very long ago, and here we are trying to second-guess a civilization perhaps billions of years more advanced, which is like one ant asking another how human beings could possibly fly through the air. Also, we don't even know for sure that they're intentionally broadcasting these signals. An automobile with poor spark-suppression in its engine produces loud, pulsarlike signals when it passes by your radio. Who knows? Maybe we're receiving the noise from their power transmission system—or from their multi-quark zartwortlers, a gadget that won't be invented on Earth for another ninety-two centuries.

There is one man who has been thinking of extrapolations thousands of years into the future. He's Princeton's Freeman Dyson, who's probably the Earth's first inhabitant to fully deserve the title of "Cosmic Architect." Way back in the prepulsarian days of 1959, he suggested that very advanced civilizations, bound only by the presently-known laws of physics, may surround their parent star with spherical shells made from dismantled planets.

Such a technologically advanced civilization would, he suspects, redesign its planetary system to utilize the energy from the biggest thermonuclear reactor around—a star. These creatures could build a shell around the star, capture all of the stellar energy, and expand their civilization over it. Such a shell—known in the trade as a Dyson Sphere—could provide a comfortable home for a billion trillion people (a comfort for those who aren't optimistic about the population explosion). These shells might block our view of the star, but the shell would "glow" in the invisible longer wavelengths of infrared light. If pulsars were located within Dyson Spheres, this would nicely explain why optical astronomers couldn't see them.

Is there any way to tell how big pulsars are? The pulses give us a clue. They last about a hundredth of a second, which sets a limit on the pulsar's size, thanks to Einstein's discovery that no signal can travel faster than light. If pulsar signals arise on or within some natural object, the signals from the nearest part of the object will arrive at Earth slightly before those from a more distant part. Since light travels at a couple hundred thousand miles per second, a pulse lasting a hundredth of a second must be emitted by something smaller than a couple thousand miles in diameter. If it were any larger, the pulse would last longer. This suggests that we're dealing with objects no larger than Earth, and not, for instance, something the size of the Sun. Dyson Spheres they're not.

Twinkle, Twinkle, Little Pulsar

So far, we've only scratched the surface of the pulsar. The size and distance of an object aren't enough to satisfy an astronomer. There's always something more to be learned, so the careful astronomer next asks what's left to measure.

A most important characteristic of any astronomical object is its spectrum, the picture of how its signal strength varies with frequency. Your local radio station has a spectrum in which most of the energy is concentrated about a single, sharply-defined frequency, a narrowband signal. If it isn't, it will interfere with other stations. On the other hand, most astronomical radio sources produce signals over a large, continuous band of frequencies.

The trouble with measuring a pulsar spectrum is that its intensity varies erratically. What the radio astronomers finally did was to record the intensity of the strongest pulse at each frequency. When they plotted this on a graph, it yielded a fairly straight line showing the strongest signals in the short-wave band, and the weakest at microwave radar frequencies. Such a spectrum is the signature of many natural radio-emitting objects, immediately suggesting that the pulsars may also be natural.

In fact, this analysis presented one of the best anti-LGM arguments around, because it showed that not only is the pulsar spectrum similar to that of many natural sources, but the signals are strongest in one of the worst possible bands for interstellar communication, the shortwave band, where radio noise from the Milky Way is strongest. So strong is this noise that it wipes out the pulsar signals we receive at the lowest frequencies, and as Frank Drake pointed out, "This says that if this is an intelligent signal, it's coming from a stupid civilization."

Nevertheless, in a talk on pulsars given at Cornell, astrophysicist Thomas Gold noted a flaw in such reasoning. Although he believed the LGM theory unlikely, he pointed out that if aliens were responsible for the signals, they might simply be controlling some natural radio-emitting phenomenon. They might be taking advantage of a natural object like those we regularly observe with our radio-telescopes, just as a person with a mirror can send coded messages for many miles by effortlessly using a mirror to divert some of our Sun's enormous light energy.

Another objection to the LGM theory was that sending signals in radio pulses spread over the entire shortwave band is a stupid way to transmit information. The smart way to do it is to concentrate the signal into narrow bands of frequencies, the way human radio stations do.

Still, apart from the possibility that it may have been too inconvenient or expensive to do any other way, the little green men could have been smarter than we gave them credit for, as there's a peculiar facet of the pulsar signals tough to explain by a natural process, but which has a nice interpretation if the signals are from little green disc jockeys. On top of that, the LGM interpretation of this pulsar peculiarity incidentally demonstrates that the little green men might not have been stupidly transmitting a broadband signal. It only *looked* that way at first because what they could have been doing was broadcasting lots of independent signals on many frequencies simultaneously, just as we on Earth broadcast simultaneously on thousands of independent stations.

This peculiar facet of the pulsar signals is seen when you look at the same pulse on different frequencies, allowing for the delayed arrival of the pulse at lower frequencies. You find the amplitude at one frequency completely unrelated to that at any other: sometimes the pulse will be intense on one frequency and weak on another, sometimes vice versa, and sometimes they're both the same.

It was difficult for theoreticians to figure out what could cause a natural radio source to put out pulses spaced so regularly, yet varying with that peculiar randomness, but if the transmissions were those of extraterrestrials, the answer could have been simple: we might have been tuning in different channels. One channel could be broadcasting Lesson #27,681 of "Medieval Pulsar History"; a second, "How to Protect Your Civilization from Destruction by Nuclear and Other Primitive Weapons"; channel 138 might be continually rerunning "An Introduction to the Little Green Language," to help us understand the other broadcasts; and another might carry reruns of last century's big hit series, "Pulsar Trek."

"Surely the nature of the signals is clear," wrote Queens College (New York) physicist Banesh Hoffmann in a letter to the *New York Times*. He noted that at some frequencies the pulses fade out for about three minutes out of every four:

> A simple extrapolation from what we have already achieved on Earth shows that the radio astronomers have detected an advanced civilization by its television transmission. The quiescent periods, too faint to detect, correspond to the programs, and the precisely spaced pulses to the bunched commercials. The pattern of three minutes of programs followed by a minute devoted to some forty loud commercials shows that the civilization is indeed more advanced than our own. Yet we need not hang our heads in shame: We are clearly not far behind.

Professor Hoffmann notwithstanding, a Herblock cartoon observed that most scientists felt the signals did not include commercials. He perceived that this fact alone proved conclusively that pulsars were from extremely advanced civilizations.

The explanation for this phenomenon eventually turned out to be a side effect caused by the interstellar plasma. Blobs of plasma interfere with the pulsars' radio signals, exactly as blobs of gas in our atmosphere interfere with starlight, causing twinkling. The strange, random fading of pulsar signals is nothing but interstellar twinkling.

The Answer

Such, then, was the reasoning scientists wrestled with to try to rule out the possibility that pulsars were intelligent signals. What turned out to be more fruitful was to try to imagine what natural object could produce such radio pulses.

As soon as pulsars arrived on the scene, everybody dreamed up every theory they could, but most of them were quickly shot down by the new facts the radio astronomers kept uncovering. "About once a day," said Frank Drake in 1968, "one of these theories shows up in the mail. It's been suggested that we set up a system whereby anyone presenting a theory has to pay a fee of ten dollars for the honor of so doing, and when, in the end, we find out which theory is correct, half the money goes to the bright person who suggested it, and we keep the rest to run the observatory."

When the dust had settled and the flakiest theories had been discarded, three non-LGM theories survived. The first was proposed in the same historic paper in which the British announced their discovery of the pulsars. They speculated that the signals might result from pulsations of fantastically compressed forms of matter known as *white dwarf* and *neutron* stars. White dwarfs are stars that have about the same amount of matter as in our Sun, but they've evolved to the point where they've collapsed into a volume the size of the Earth, so that a piece of white dwarf the size of a flea would weigh about a ton.

A neutron star is what might be left after a star explodes. It would be so dense as to make a white dwarf look as substantial as a ping-pong ball—the same amount of matter is crushed into a volume only a few miles in diameter. A neutron-star flea would weigh a hefty million tons. White dwarfs have actually been observed by telescopes, but neutron stars still dwell in the theoretician's imagination.

The idea of the Hewish group, then, was to imagine you had such a star that was pulsating like a heart. What makes life difficult for the pulsating-star advocates is that white dwarfs should have pulsation periods of more than two seconds, whereas neutron stars should pulse more rapidly than even the fastest of Hewish's pulsars—which puts the pulsars right in the one place where neither explanation works.

How do you go about dreaming up a pulsar theory? You consider every conceivable kind of astronomical object, and try to figure out a way to get a rhythm of about one beat per second out of it. One thing that might do it is a binary star system, because if two stars are close enough, they'll whip around each other every second or so. The only kind of stars that fill the bill, though, are our old hypothetical oddballs, the neutron stars. Other stars are too big—they can't get close enough to do the job. The neutron stars would have a strong magnetic field between them that would accelerate charged particles ejected from the stars, creating the radiation we detect. Unfortunately, this theory has the rather embarrassing feature that the system should rapidly collapse.

What plagued the binary boys was that Einstein's general theory of relativity predicts that binary stars ought to radiate gravitational waves the way moving electrons emit electromagnetic waves (such as light and radio waves). These gravitational waves would carry away the energy of the stars, and the neutron-star binary would collapse in a few years, so the pulses could never remain as fantastically constant as we find them to be.

The third theory was the one that looked the most promising. This was Thomas Gold's.

On first hearing Gold's precise, flawless British speech, you're sure he must hail from Oxford. Actually, it was Cambridge, and if you listen carefully, you may even detect the faint accent of his native Vienna. He's somehow found time to devise major theories of the Earth's magnetic field, the lunar surface and cosmology, in addition to a fascinating theory that the Earth is full of natural gas, which if right, could solve all of humanity's energy problems for centuries. He's a busy man.

What else can you do to a neutron star to make it into a pulsar other than vibrate it or revolve two of them around one another? You can spin it. A neutron star would be spinning very fast if it had formed from a rotating star, and Gold figured it was reasonable to expect some neutron stars to be spinning at about one revolution per second. He offered this spinning-star pulsar model as the reason the pulse rate's so remarkably constant—it's awfully hard to change the speed of a massive flywheel like the Earth or a star. This constancy is the most important clue. It is difficult to imagine anything but a flywheel that could so constantly produce pulses of that speed.

The problem was then to explain why such a spinning star should radiate.

Gold proposed that the neutron star might eject a plasma of charged particles much as our Sun often does. The neutron star would have an extremely intense magnetic field "frozen" into it— extending well into space, as does Earth's—which would whip the particles up to almost the speed of light, where they'd radiate. As the pulsar rotated, its radio signals would sweep the heavens like the beam from a cosmic lighthouse beacon.

In 1968, trying to salvage the LGM theory, I asked him this: suppose an alien civilization found a neutron star nearby doing nothing more conspicuous than spinning once a second. Might not they try to convert the star into a radio transmitter? "Yes, that's a very nice idea," said Gold. "Blow up little atomic bombs, and as they blow up, they supply plasma for a brief moment, get accelerated, and radiate. With a neutron star sitting right next to us, I'm sure I'd propose that we should do that."

But he was very skeptical about the LGM theory. If an alien community were to broadcast that amount of energy, Gold pointed out, how could they be so blasé about contacting new civilizations that they would not offer a clue about themselves in that transmission?

"Suppose," Gold mused, "somebody here has given me the job to put out some *huge* amounts of radio energy which I know would go to millions of other stars in a detectable way. Then surely I would be so tempted to put in—if I was allowed to do so—some kind of clue that this is not a natural, but an artificial, signal. And it would have been so easy to include such a thing. Either these boys just absolutely don't care a damn—and that I find very hard to believe . . . I mean, we're interested in the ants!"

What the public sometimes fails to understand is that the ultimate proof of a scientific theory is not how good it looks, or how respectable its believers are, or even whether it explains past observations. The proof is in its ability to predict. Gold's theory faced its first major test when the National Radio Astronomy Observatory discovered a pulsar in the Crab Nebula, the remnant of a star whose explosion was seen by the Chinese on July 4, 1054.

The Crab pulsar was the fastest one yet, pulsing every 0.033 seconds. The great Arecibo radiotelescope was aimed at it. Although Gold's theory predicted that a neutron star should spin with fabulous constancy, it also predicted that it should slow down by an extremely small amount—an amount so small that it had not been possible to measure yet. The enormous energy radiated by the pulsar should act like friction, causing it to slow down like a toy top.

His theory predicted that a young pulsar—one spinning rapidly—would slow down the most quickly. The Crab pulsar was spinning so fast it should slow down noticeably within a few days. If it didn't slow down soon, the theory would be in big trouble, and would probably have to join numerous other great ideas in the junkyard of bad pulsar theories.

But in a few days, the Crab pulsar became the first one to demonstrate a measurable change in its pulsing rate—just as Gold's theory had predicted! His spinning neutron-star theory of pulsars is now the accepted explanation.

The Crab pulsar was also the cause of some sadness to a few astronomers, because a year or so before the discovery of pulsars, researchers at a major radiotelescope had planned to survey the Crab with some fast-response receiving equipment. To their eternal sorrow, they canceled the project, and thereby forfeited their chance to become the discoverers of one of the biggest astronomical bombshells of all time: pulsars!

John Ziman, in the preface to his text *Principles of the Theory of Solids*, presents an astute observation on the course of modern science that summarizes nicely the lesson of pulsars:

Today's discovery will be tomorrow part of the mental furniture of every course of graduate lectures. Within the month there will be a clamor to have it in the undergraduate curriculum. Next year, I do believe, it will seem so commonplace that it may be assumed to be known by every schoolboy.

The Lesson of the LGM's

Even if pulsars are not extraterrestrials, they are one of the most important discoveries in the history of astronomy. Physicists since Einstein have been striving to measure tiny deviations from the old theories that concepts such as relativity predict. If Einstein was right, the very fabric of space and time is warped by the presence of gravity, but the effect is tiny near a humdrum star like our Sun. At last, neutron stars presented natural laboratories in which we could begin to test those theories under conditions which are not just marginal but *overwhelming*. Strange, predicted phenomena such as gravitational waves—waves similar to those in the neutron-star-binary pulsar theory—could now be tested.

Moreover, the existence of these outlandish objects made the concept of an even more outrageous phenomenon, *black holes*, almost inescapable. Black holes are stars that are a bit heavier than neutron stars and so massive that their gravity would prevent even light from escaping. Space and time would be distorted to the breaking point, creating pockets of the universe where the known laws of nature might completely break down.

Eventually, a number of probable black holes were indeed detected. So far, Einstein seems to have been right.

It is not surprising, then, that in 1974, the director of the pulsar project, Antony Hewish, won the Nobel prize in physics. What surprised a lot of people was that the woman who made the actual discovery, Jocelyn Bell (now Burnell), did not share the award.

Another radio astronomer's false alarm. Radio signals normally emitted by the nebula W49 are shown by the bell-shaped curve in (a); this plot shows the strength of an emission feature centered at a frequency of 7792 megahertz, produced by electrons within some of the nebula's hydrogen atoms. A peculiar observation produced the results in (b). When the adjacent data points are connected, as in (c), the plot takes on the appearance of intense interference. But when the observed data points are connected more imaginatively, as in (d), a seemingly startling pattern emerges.

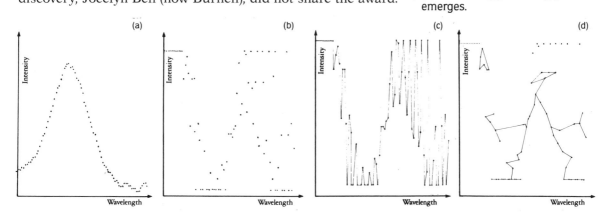

*Little Green Men
and Nobel Prizes*

An outcry was heard from those who thought this was sexism in the extreme. The Nobel Committee claimed that sex had nothing to do with its decision. It is their longstanding policy to award the prize on the basis of a body of work, not necessarily a particular single discovery, although usually any scientist receiving the honor has made at least one singular discovery that makes him stand out from the crowd of his fellow hard-working, but more routine, colleagues. Hewish had conceived and overseen the construction of the radiotelescope that made the discovery, and had insured that the proper tests of the properties of the new objects were made. A graduate student working on a project who happens to make a great discovery is often assumed to have been just an assistant.

The final word must be from Jocelyn Bell Burnell. She graciously agreed with the Nobel Committee that Hewish deserved the prize and that she did not. Still, it seems unfair. But her discovery certainly did not hurt her career. She received her well-earned Ph.D. and is today an astronomer at the Royal Observatory, Edinburgh, Scotland.

The "Wow" signal, received at Ohio State University in 1977. Numbers represent the normal background noise of the sky on this computer printout; encircled letters show an unusually strong signal. Because it didn't repeat, the signal can't be identified. (Courtesy Ohio State University Radio Observatory.)

The story of the pulsars, with its false trails and complex clues, probably reflects the difficulties we will face if SETI succeeds. It also provides an important lesson to critics of SETI: even if no extraterrestrials are found, we are almost certain to find new phenomena that will improve our understanding of the universe.

The "Wow" Signal

One last type of false alarm is worth mentioning to show another way the universe conspires to frustrate SETI researchers.

Once in a while, radio astronomers receive mysterious signals whose origins, to this day, are unknown. At Ohio State University, where the long-suffering SETI scientists John Kraus and Robert Dixon have kept their project running despite close encounters with golf-course developers, they once received a strong signal of unknown origin. The observer was so impressed with it that he wrote "WOW!" on the chart, and it is now known as the "Wow" signal.

The problem with such signals is that if they do not repeat, we cannot perform tests on them. We cannot be sure whether they are from space or from some earthly interference. And if they are from space, are they produced by some previously undetected natural phenomenon? Or are they the noise of some unimaginable event that took place in an alien civilization centuries ago?

Astronomers occasionally search the sky again at those places and frequencies, hoping the signal will repeat. . . .

A fan of the now-defunct SETI magazine formerly published by Kraus and Dixon. (Jim Arthur in *Cosmic Search*.)

10

Darth Vader vs. E.T.

THE KILLING OF SETI

If we continue to allow NASA to pursue this effort to intercept signals from some hypothetical intelligent civilization, we are sending exactly the wrong signal to the American taxpayer.
Senator William Proxmire (1981)

IT WILL NOT SURPRISE YOU to learn that many people at NASA are extremely interested in SETI. Several times, they have proposed building a sophisticated, special purpose, computerized receiver that would be dedicated to SETI. It would listen to millions of channels simultaneously. But each time, until recently, the NASA effort was killed. In 1981, for example, less than three months after the launch of the first Space Shuttle, NASA's SETI program was axed from the budget by one of the most extraordinary amendments in the history of Congress.

One senator objected to the proposed SETI project. NASA wanted to spend two million dollars per year on it. That's *million*, not billion! That amount is so small that it does not ordinarily appear in Congressional budgets. But the senator thought this was a waste of the taxpayer's money, so he had the following amendment added to the federal appropriations bill:

> . . . none of these funds shall be used to support the definition and development of techniques to analyze extraterrestrial radio signals for patterns that may be generated by intelligent sources.

Senator Proxmire.

The senator was William Proxmire of Wisconsin, best known perhaps for his Golden Fleece Awards to those military and civilian projects he regards as particularly wasteful. He has often earned the ire of the scientific community by attacking research projects,

145

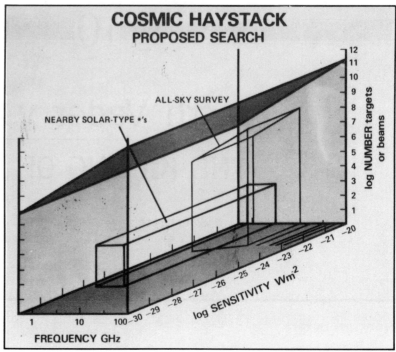

COSMIC HAYSTACK
PROPOSED SEARCH

The proposed NASA search of the cosmic haystack. NASA wanted to cover far more of the haystack than had ever been done. The Ames Research Center would concentrate on nearby Sunlike stars; JPL would cover the entire sky, but at less sensitivity.

and had previously given NASA a Golden Fleece Award for SETI, which helped kill the project in the 1979 fiscal-year budget. His 1981 amendment ensured that, in the eyes of the SETI community, he was now Darth Vader.

Interestingly, two congressmen spoke out against this amendment: Representative C. W. Young of Florida and Senator Harrison Schmitt of New Mexico. Senator Schmitt's position is of particular interest, as he is the only senator to have walked on another world. Senator "Jack" Schmitt was one of the two last men on the Moon, in *Apollo 17*, and the only scientist to go there.

This was an interesting scene in Washington. On the one hand, we had a senator arguing, in effect, that the idea of searching for intelligent life on other worlds was science-fiction nonsense. On the other hand, the senator he was debating was an astronaut who had been to the Moon—an accomplishment that just two decades earlier was regarded as science-fiction nonsense.

Unfortunately, the amendment was passed by Congress. However, the year before, Carl Sagan and Bruce Murray (then Director of the Jet Propulsion Laboratory) conceived an organization that would be a private, nonprofit group dedicated to advancing the cause of planetary exploration and SETI. They called it The Planetary Society, and it became the country's fastest-growing public-membership organization, rising to more than 100,000 members—the largest private pro-space group in the world.

I became their SETI Coordinator, and as a result was at a SETI meeting at NASA two weeks after the Proxmire amendment passed. The NASA SETI program was a project of the two California NASA centers, the Ames Research Center in northern California, and JPL in Pasadena. This meeting was at Ames.

The amendment was scheduled to take effect in one week, on October 1, 1981. We were seated around an appropriately coffin-shaped table, and the mood was as grim as if Darth Vader were about to unleash the Death Star on our planet. The mood was broken by a moment of laughter when we learned that the *Washington Post* had printed an article criticizing the killing of SETI. The title referred to *Mork and Mindy*, the popular television comedy about the alien, Mork, and his visit to Earth. The headline read: "Are You There, Mork? Please Reply By October 1."

We tried to extract some optimism from the report that Proxmire had said that if SETI is so important, NASA could come back and try again the next year. But to do this would require the approval of NASA Headquarters in Washington, which might write off SETI as a lost cause.

The next morning, we were gathered around the table again. NASA's SETI Program Manager, John Billingham, entered the room.

One of NASA's antennas that may be used for SETI. At 85 feet in diameter, it is one of their Goldstone antennas on the southern California desert.

"I'm sorry, Congressman Sledge, this passenger section was personally designed by Senator William Proxmire."

He had just gotten off the phone with Headquarters, and he was smiling. In his eloquent British voice, he told us that Headquarters had decided to do just what the senator suggested. SETI would have another chance.

The mood brightened.

The SETI meeting presented an opportunity for The Planetary Society to act. The Society was not rich enough to finance NASA's SETI research, but it could pay for smaller SETI programs.

E.T.: Phone Harvard

One of the scientists present was a young Harvard physicist named Paul Horowitz (no relation to the *Viking* project's Norman Horowitz). He presented a proposal to search for a special type of signal that another civilization might broadcast as a beacon, to simplify the search by an emerging civilization such as ours. By concentrating on just this type of signal, he devised on paper a receiver so compact that he called it "Suitcase SETI."

What he proposed was to look just at magic frequencies, extending the original idea of Cocconi and Morrison to include not just the hydrogen frequency but also the frequencies of the water and hydroxyl (OH) molecules prominent throughout our galaxy.

If you were to tune in to one of these magic frequencies with a receiver, you'd find that Mother Nature is a sloppy radio broadcaster. The hydrogen signals are smeared out over an appreciable part of the dial, or in *broadband* signals. Horowitz proposed to look for signals that were broadcast at extremely precise frequencies—*ultranarrowband* signals—making them very different from natural

Suitcase SETI, the system tested by Paul Horowitz at Arecibo. The main instrument is at bottom left, the device which analyzes 131,000 channels. On top of it is a videocassette recorder used to store data, and at right is the Wicat microcomputer that controls the system.

ones. If he found such a signal, it would probably be artificial.

This design would avoid many sources of interference on Earth and in space. He proposed using a computer to search 65,000 channels, each channel being 0.015 Hertz (cycles per second) wide. The bandwidth of 0.015 Hertz, thousands of times smaller than natural signals, was chosen because signals narrower than that would be broadened by the effects of ionized interstellar gas. Atoms and molecules typically have bandwidths of thousands of hertz at these frequencies, so a signal with all its energy concentrated into a fraction of a hertz would probably stand out from the noise.

So impressive was the Horowitz presentation at this conference that the scientists voted on the spot to endorse the design. The Planetary Society then discussed it, and within two weeks of his presentation, Horowitz was given the go-ahead by the Society. This was lightning speed in the annals of scientific funding, where normally it takes about a year for a proposal to be accepted.

Horowitz and his collaborators, Ivan Linscott, Cal Teague, Kok Chen and Peter Backus, built Suitcase SETI at Stanford University, with help from people at Ames. About six months after the proposal, Horowitz took it to Puerto Rico, where he attached it to the Arecibo radiotelescope, using the thousand-foot dish.

They ran Suitcase SETI off the giant antenna and listened to more than two hundred of the nearest Sunlike stars. No artificial extraterrestrial signals were found, but of course this was just the beginning.

Luke Skywalker vs. Darth Vader

After the anti-SETI amendment, The Planetary Society began considering what to do to get Congress to reconsider the program. The most important move was to arrange for Society President Sagan to meet with Senator Proxmire. Two years later, at the first SETI Symposium of the International Astronomical Union, Sagan described what happened.

Sagan had originally become acquainted with Proxmire in a different context, in concern about the threat of nuclear war and other issues. Sagan had, for example, once written him a letter about why some construction of the Jet Propulsion Laboratory—which Proxmire was opposing—was important. The senator had written back saying that Sagan's explanation was good, so he'd now support it, and he did. Through this incident, Sagan got the sense that he was a reasonable person and that it was worth trying to talk to him about SETI.

This time, when Sagan visited Proxmire, the senator seemed

preoccupied with the nuclear war issue, so they talked for a while about that. When Sagan did bring up the SETI question, Proxmire said, "The whole idea is nonsense." His reasoning was that if extraterrestrial intelligence existed, it would be here by now, and it would have been discovered. This is a controversial argument which had been published in the scientific literature. So what Sagan pointed out is that it's very likely the Earth has not yet been discovered by other civilizations.

He then described the Drake Equation, emphasizing its last factor, the *lifetime* of a technical civilization: if the civilization is short-lived, the number of them in the galaxy is few, and the average separation between them is very large—thousands of light-years or more—so that the likelihood of our being discovered is actually quite small. Sagan told Proxmire that even if the lifetimes of technical civilizations are long, the galaxy is sufficiently empty that the nearest civilization isn't all that close, maybe hundreds of light-years away.

"Well," Sagan said, "he had never heard this! He had never heard *any* argument about the separation of the stars, about *any* aspect of this, but especially, he had never heard of the connection between the longevity of advanced civilizations and the question of how many of them there are. And before my eyes, I could see two neural nets

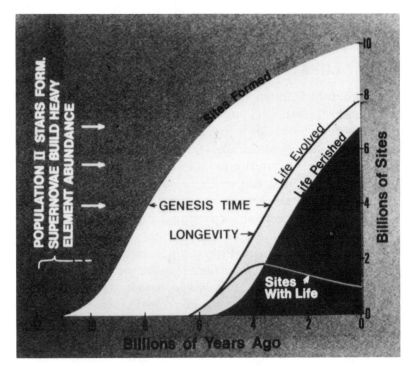

A theory of the life and death of planetary civilizations. As the universe ages, the heavier atoms of life are formed from hydrogen. Civilizations arise; some die. "Sites with life" shows the resultant curve of surviving cultures.

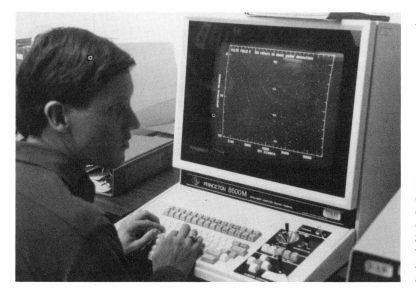

Blind scientist Kent Cullers at the controls of a Stanford University computer, demonstrating a program he developed for the NASA SETI project. The program searches for radio pulses, a possible sign of another civilization. It will also automatically discover pulsars.

which were in different parts of his brain that had never met, be introduced to each other.

"And he then said to me, 'Do you mean that if we find some evidence of extraterrestrial intelligence, that will mean that somebody elsewhere has avoided self-destruction?' And that's very much what I think many of us believe. I said that was likely, and he said he would go away and think about this problem. Well, he thought about it, and he withdrew his objection to funding SETI." (Sagan emphasizes that his visit was not the only factor in educating the senator—other scientists and citizens also played an important role.)

That the possibility of World War III occurs in a scientific equation is one of the most remarkable facets of SETI, and one which has fascinated a good many thinkers in the search for extraterrestrial intelligence. To some, the mere hope that would result if we found that other civilizations had survived the kind of global dangers we now face is sufficient reason to support SETI.

Sagan drew two lessons from this encounter: (1) There is a failure of the scientific community to explain to our representatives the fundamentals of SETI, such as the Drake equation; and (2) the connection between the search for extraterrestrial intelligence and the question of the self-destruction of civilizations is a powerful and compelling issue.

But the battle was not over. SETI was now in the official, proposed federal budget, but it still had to get through the rest of Congress.

By a twist of cosmic irony, the movie *E.T.* was released at this time, and it set box-office records. In fact, it grossed more than three million dollars a day for the first month of its release—more money per day than NASA was asking to search for real extraterrestrials for a year.

Congress debated the national budget with the speed of a drifting continent. Months passed. SETI scientists waited in suspense as budgetary meteors threatened to shoot down their project.

One such meteor was the balanced-budget amendment. Designed to fight the $100 billion-plus expected federal deficit, its passage would have meant cutting many federal programs, and only those programs with large constituencies seemed likely to survive. The SETI scientists knew only too well that extraterrestrials do not have large constituencies, at least not on this planet. Then, too, it was an election year, and there was endless debate and horse trading about practically everything else in the budget. No one knew whether SETI might get traded for someone else's horse.

September 30, 1982. The new fiscal year was approaching. Congress knew it had to pass some bills or the government would go bankrupt. However, something mightier than a meteor shower intervened: Party time! Both Republicans and Democrats had parties scheduled that night. So they adjourned without passing the bills.

October 1 was a busy day. First the government went bankrupt. Then Congress passed an interim funding bill, killed the balanced-budget amendment and approved only one of the thirteen budgetary bills that the government runs on. The one approved budget was for Housing and Urban Development and Independent Agencies. Fortunately, NASA is an Independent Agency.

NASA was in the SETI business.

11

The Russians Are Looking, the Russians Are Looking!

INTERNATIONAL SETI

Don't hurry to reply, but hurry to listen.
Old Russian proverb

MORE THAN FORTY SETI searches have been carried out since Project Ozma, using a variety of different receivers and antennas in Canada, France, West Germany, Holland, Australia and the Soviet Union, as well as the United States. But the number of searches is misleading. Only a tiny percentage of the possible frequencies, positions and transmission modes have been searched. The Cosmic Haystack has barely been touched.

The strongest interest in SETI has consistently been in the Soviet Union, where more searches have been carried out than in any country but the United States. Still, the Russian attitude toward SETI often contrasts in unexpected ways with that in the West. Sometimes these two societies seem as alien as if they were on different planets.

Russian interest in space exploration goes back at least to the nineteenth century. They had important astronomical observatories, and they had a man whose ideas would cross the centuries, span the Revolution, and give the world *Sputnik*. His name was Konstantin Tsiolkovsky.

He was born the son of a forester, in 1857, two years after the birth of Percival Lowell and two years before the publication of Darwin's *Origin of Species*. When Tsiolkovsky was ten, scarlet fever made him

"Although humans make sounds with their mouths and occasionally look at each other, there is no solid evidence that they actually communicate among themselves."

153

deaf, which prevented him from entering school but did not handi-cap his mind.

He eventually became obsessed with the idea of space travel, and did many experiments and calculations that laid the foundations of the technology. By 1897, while horses were still most people's mode of transportation, he had already derived the basic equation of rocket motion, which the Russians call the Tsiolkovsky formula. He built the first Russian wind tunnel. He proposed the radical idea of a rocket fueled by liquid hydrogen and oxygen, the major fuel of today's Space Shuttle—and he did this in 1902, the year before the Wright Brothers' first flight!

He expressed many of his visionary ideas in science-fiction stories about the conquest of the cosmos. He was convinced that humani-ty would colonize the solar system. In 1911, he wrote:

> At first we inevitably have an idea, a fantasy, a fairy tale, and then come scientific calculations; finally execution crowns the thought. My work has to do with the middle phase of creativity. More than anyone else I am aware of the chasm that separates an idea from its accomplishment, for during my whole life I not only did many calcula-tions but also worked with my hands. But there must be an idea: ex-ecution must be preceded by an idea, precise calculation by fantasy.

For this, he was regarded as a crank by many, but not by all. He had friends on the winning side of the Russian Revolution of 1917, and received many honors following that event. His ideas were ex-panded upon by the German and American rocket pioneers Her-mann Oberth and Robert Goddard, eventually giving birth to the jet and rocket programs of those countries. He died in 1935, less than a quarter century before *Sputnik* made his dream start to come true.

Most western space experts regard Tsiolkovsky as one of the great pioneers, but few realize the power that his ideas have exerted in the Soviet Union. His vision, through both his scientific writings and his science fiction, has inspired generations of Soviets and helped sustain the strong pursuit of rocketry and space explora-tion. They have maintained their dedication to space exploration consistently for decades, while America's passion has alternately flamed up and died down, usually in erratic response to Russian successes and failures in space.

In 1911, Tsiolkovsky wrote these prophetic words:

> Mankind will not remain on the Earth forever, but in the pursuit of light and space will at first timidly penetrate beyond the limits of the atmosphere, and then will conquer all the space around the Sun.

He even captured the spirit of SETI in his essay, *Living Beings in the Cosmos:*

Can communication be accomplished between neighboring suns? . . .

The epochs that have become lost in the infinity of time produced beings that achieved perfection just as beings made up of "our" matter are achieving it . . .

Our imagination presents to us an infinite number of epochs in the past and in the future, each with its living beings. What are these beings like, is there any connection between them, how do they manifest themselves, can they manifest themselves, do they disappear with the arrival of a new epoch?

To Russia with Difficulty

Senator Proxmire's anti-SETI amendment of 1981 heightened the contrast between American and Soviet attitudes. That same year, the Soviets held an international SETI conference. Many of the American scientists could not attend, because the Proxmire amendment prevented the use of NASA money for travel on SETI business. So The Planetary Society stepped in with some funds and arranged for the Sloan Foundation to sponsor others. As a result, ten people

RATAN-600, a Soviet radiotelescope used for SETI.

from the United States—only one-third of those invited—were able to attend. When they arrived at Tallinn, Estonia, in the USSR, where the conference was held, they were treated almost like heads of state. They were escorted by police cars with flashing lights that stopped all other traffic. They met with four cosmonauts and a member of the Supreme Soviet. They were inundated by press and television interviews.

In the United States, hardly an article appeared.

A role-reversal of sorts occurred with the first SETI symposium of the prestigious International Astronomical Union. It was held in Boston in 1984, organized by the Greek-born Boston University professor Michael Papagiannis. This time, it was the Soviets who couldn't come. Apparently, this was a spinoff of the Russian retaliation for the Western boycott of the Olympic games in the Soviet Union.

A brighter note was provided unexpectedly by Halley's Comet. What does that have to do with SETI? Well, the Russians sent two probes called *VEGA* to Venus. Part of the *VEGA*s went on to Halley's Comet. Since the Russians wanted NASA to track the probes, and since the two countries use different frequencies on their spacecraft transmitters, NASA had to modify their Deep Space Network antennas to pick up the Russian signals. This is just what NASA SETI scientists wanted: to receive as many frequencies as possible. This was a step toward their goal of covering the entire microwave radio band. Halley's Comet, proclaimed throughout history as a bringer of misery to emperors and countries, had brought blessings to SETI.

Strangely, since the anti-SETI amendment, the positions of the two countries have nearly reversed on SETI. While the United States has speeded up its SETI work, the Soviets have slowed down theirs.

Frank Drake has had numerous contacts with Soviet SETI researchers, so I asked him what the Russians are doing now. According to him, the Russians have two basic programs. One is a 230-foot diameter, very high-precision antenna that they are building near Samarkand. It is to be used for conventional radio astronomy, but will also be used in SETI searches. The antenna is a duplicate of several others that already exist and are used in their deep-space tracking-network.

The other project is at Gorky, headed by V. S. Troitsky. Troitsky believes we should look for broadband pulses, which would be very powerful, so he's building a system of hundreds of one-meter (39-inch) diameter dishes, which will be set up so that astronomers can view the whole sky at once. With the dishes that small, each will cover a large area of the heavens, so it won't take too many to cover the whole sky. Drake doesn't think this is very good logic, because they're giving up sensitivity this way.

There has often been disagreement between East and West over the best strategy for designing a search. Americans have tended to favor narrow frequency bands (like those of ordinary broadcast radio), while the Russians often prefer to look for signals spread over broad ranges of frequencies.

The friendly disagreement between these two "alien" cultures on Earth may be to everyone's advantage. There may be aspects of interstellar communication we have not yet thought of, so in my opinion, the more types of searches conducted by humanity, the better our chances of success.

Other Nations

Numerous small-scale SETI projects have been conducted in other countries, and some of them may grow to rival those of the United States and Soviet Union. Some of the institutions that have sponsored SETI programs to date are: Australia's Commonwealth Scientific and Industrial Research Organization in New South Wales; France's Observatoire de Nançay; Canada's Algonquin Radio Observatory in Ontario; the Dutch Westerbork Synthesis Radio Telescope; and the West German Max Planck Institut für Radioastronomie. SETI is becoming as international as space itself.

In addition to the professional observations of Canadian scientists at Algonquin, there is a dedicated Canadian amateur, Robert Stephens, who obtained two large (50-foot) surplus antennas and has been turning them into SETI antennas—with few funds and much difficulty—in the Yukon, where the principle sources of interference are his electric razor and passing motor boats.

One globetrotting SETI scientist, Jill Tarter of NASA's Ames Research Center, who has participated in more SETI searches than almost any other person, summarized the state of worldwide SETI in a report to the International Astronomical Union. "Although SETI is still acknowledged to be a long shot," she wrote, "and is often likened to buying a lottery ticket, the entries . . . give clear evidence of a growing commitment to pay the purchase price for that lottery ticket, instead of waiting around hoping that a ticket will fall out of some astronomer's pocket!" She went on to add, "I believe that the multiplicity and visibility of these many observational programs have helped to establish the current climate wherein NASA can be engaged in the process of preparing to conduct the first large scale systematic SETI program."

An Australian radiotelescope, part of NASA's Deep Space Network, a 210-foot diameter dish.

NASA and The Planetary Society are now exploring the possibility of conducting SETI from the Southern Hemisphere. Scientists love to quote Murphy's Law: "If anything can go wrong, it will." If Murphy's law applies to SETI, it could mean that the most detectable civilization just happens to be discoverable only from down under, where but a few SETI searches have been conducted so far.

Argentine and Australian scientists have discussed possible SETI observations with North American scientists, and it is likely that a radiotelescope will be used for this purpose in one of those countries within the next several years. A major development in that direction came at the end of 1985, when the first Latin American SETI conference was held at the University of Buenos Aires. The Society arranged for JPL to send scientist Bruce Crow to make "first contact" with his colleagues in that country.

Some Japanese scientists also hope to start their own SETI project. They have recently built a superb radiotelescope, one of the best in the world, at Nobeyama, Japan. Run by the University of Tokyo, it has a 150-foot diameter dish of exceptional precision. Some

Australian radiotelescopes used by NASA's Deep Space Network. Inset is the Parkes antenna, which was used in conjunction with the others to receive the *Voyager* signals at Uranus.

of their astronomers hope to use it, in part, for SETI observations. I expect them to set a high standard, stimulating more healthy SETI competition between nations.

And, as if to add to the international flavor of the goulash of nations involved in SETI, the next SETI symposium of the International Astronomical Union is scheduled to be held in Hungary.

In 1985, I asked Roald Sagdeev, the head of the Soviet equivalent of JPL, what they were planning for future SETI programs. He said that they are hoping to build a radiotelescope in space that would be used for both astronomy and SETI. This is an idea that has been around for quite a while in both countries, and it remains to be seen whether either can make this splendid concept a reality.

Later, the Executive Director of The Planetary Society, Louis Friedman, visited Russia and spoke to radio astronomer Nicolai Kardashev, a leading SETI theoretician. The slow progress on their new antenna at Samarkand was hinted at by a cartoon in his office: a man holds a radio antenna, while sitting on top of a turtle. He's crying.

Kardashev revived his long-time dream of a giant, one-kilometer radiotelescope in space, operating together with Earth-based telescopes. He hinted that this would be a perfect project for future Soviet-American space cooperation.

The Samarkand project and a small orbiting radiotelescope will probably be operating in the 1990s.

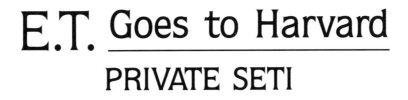

E.T. Goes to Harvard

PRIVATE SETI

Money won't buy happiness, but it will pay the salaries of a large research staff to study the problem.
Bill Vaughan

THE NEXT TWO MAJOR EVENTS in SETI turned out to be acts of God and Steven Spielberg.

After NASA's SETI program was approved, and Suitcase SETI had been built, everything went smoothly for a while. Suitcase-SETI designer Paul Horowitz took his instrument down to the great Arecibo radiotelescope and looked at more than 200 nearby Sun-like stars. He did not find any suspicious signals, but the system worked well.

He took his device to Harvard, where he is a professor of physics. There he found an 84-foot diameter radiotelescope that was about to be boarded up. Located in the small town of Harvard, Mass-achusetts, less than an hour's drive from Harvard University, it was run by the university together with the Smithsonian Astrophysical Observatory. He had originally conceived of Suitcase SETI as a system that could be carted around the world to whichever obser-vatories would give him time on their radiotelescopes, but he saw now the possibility of a permanent SETI facility.

The Planetary Society agreed and Suitcase SETI was permanently installed at Harvard, Massachusetts, paid for by the contributions of thousands of Society members who donated funds specifically for SETI.

We conducted a contest to name the new facility, and received hundreds of suggestions, including Big Ears, Clark Gable (because he had big ears), ET&T, Deep Peep, and one submitted by someone who is evidently a citizen's band radio afficionado: That's a Big 10-4, Cosmic Buddy! The winner was Project Sentinel.

When Sentinel began operations in 1983, it became the world's most advanced operational SETI receiver, and the only permanent, professional SETI observatory operating besides the Ohio State University one. Every day, Sentinel's dish would be raised half a degree. Then the antenna would remain stationary while the Earth rotated for twenty-four hours, observing a strip of sky half a degree wide. In less than a year, Sentinel would scan the entire sky visible from Massachusetts. It covered 131,000 adjacent ultra-narrowband channels, initially centered around hydrogen's "magic" frequency, 1420 MHz.

After Sentinel had been running nicely for a while, Carl Sagan and his wife and co-author, Ann Druyan, had a close encounter with Steven Spielberg in the course of a discussion on joint projects in Southern California. During that discussion, the director of *E.T.* promised to donate $100,000 to The Planetary Society, for SETI.

Paul Horowitz, too, was having stimulating discussions with his colleagues. He began to think about the possibility of increasing the number of channels in his system. 131,000 channels sounds like a lot—and it is. No other SETI system had run with so many channels, although the one NASA was designing would eventually have millions. But there is a special reason why more channels could

Paul Horowitz of Harvard University at the controls of Project META.

greatly benefit Sentinel. The reason is that everything in the universe is moving.

If an alien civilization transmits an ultra-narrowband signal of the type assumed by Horowitz, then its frequency will be changed by the same Doppler effect that reddens the light of receding galaxies. All the stars in our Milky Way galaxy orbit around the Galactic Center much as planets orbit the Sun, but the orbits of stars are not so neat. With several hundred billion stars in our galaxy, the stellar traffic is, at first glance, as congested as the streets of any big city at rush-hour. The orbits cross randomly. Only the vast distances of space make collisions extremely rare.

If you look at any star in the sky, the chances are that it is moving with respect to us. It may be heading more or less toward or away from us. (It will also probably be moving sideways, but that does not cause a significant Doppler shift.) If a beacon star happens to be heading toward us, then its frequency is raised, shifting its signal to a higher channel. If it is moving away, its frequency is shifted to a lower one.

Because of the limited number of channels in Sentinel, Horowitz operated on the assumption that another planet's civilization would beam its signal directly at the Sun, artificially tuning its frequency to compensate for the Doppler shift. Sentinel then compensates for the other Doppler shifts due to the motion of the Earth around the Sun and the spin of our planet.

But perhaps the aliens are not so thoughtful. Perhaps they assume we are at least smart enough to search over enough frequencies so that they do not have to compensate for our personal motion, just as a lighthouse does not shine its beam only in the direction of a ship. They might, for example, tune their transmitter to broadcast at the magic frequency of hydrogen in the center of the galaxy, which would be a logical standard frequency for all beacons in the Milky Way. If this is true, then Sentinel might miss a beacon entirely.

What would be ideal, Horowitz realized, would be a receiver with 64 times as many channels, or eight million! This would also fit in beautifully with a new idea of SETI pioneer Philip Morrison.

Remember Penzias and Wilson, the men from Ma Bell who discovered the cosmic radio noise that is the echo of the Big Bang? Well, other researchers had found that there are tiny variations in the frequency of that noise, due to the Doppler shift caused by the motion of our galaxy through the universe as a whole. Just as all the stars move around with respect to each other, the galaxies do, too. But now, this cosmic Doppler shift allows us for the first time ever to measure our motion with respect to the whole universe.

This means that an alien civilization could tune its transmitter to broadcast not at some randomly shifted frequency due to its motion, but to the most fundamental frequency of the universe: the frequency that hydrogen would have if sitting still while everything else in the universe moved around randomly. As Philip Morrison so poetically puts it, "We are boats floating in a current-washed sea, and we move, and that changes the frequencies."

Eight million channels would be enough to encompass the uncertainties in the measurement of our motion in the "cosmic rest frame." Electronics whiz John Forster, who had helped build Sentinel, suggested a way this unprecedented number of channels could be built economically, spending far less than the millions of dollars it would have cost a decade earlier. It would cost about $100,000, the amount Spielberg had promised to donate.

Horowitz could now become the first human to deliberately tune to the magic frequency of the entire universe.

Acts of God and Man

Horowitz busily set about building the new equipment with the help of colleagues, students and volunteers. So powerful would the system be that it demanded a new name, and Sentinel became META: Megachannel Extraterrestrial Assay (pronounced "metta").

In 1985, we set September 29 as the date for the official ceremony

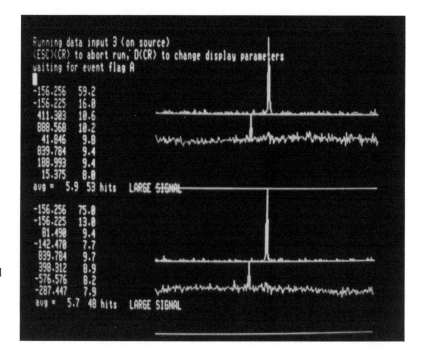

A simulated SETI signal. If a signal from another civilization is detected, it may show up as a spike standing out above the erratic background noise on this computer screen.

A schoolchildren's gift on the wall at Project META.

turning on META. We invited Spielberg to throw the switch, and he agreed.

As the day approached, Horowitz and programmer Brian Matthews frantically assembled the final pieces of hardware in Massachusetts, including some critical equipment from Ivan Linscott of the NASA/Stanford SETI project. The personnel at The Planetary Society on the other side of the continent organized the ceremony.

A couple of days before I was scheduled to leave California for Massachusetts, I checked with Lyn McAfee, the Society official who did most of the work for the ceremony. I asked her how everything was. "Fine," she replied, "except for the hurricane."

"Hurricane?" I said. "What hurricane?"

"There's a hurricane moving in to Massachusetts now."

"Terrific," I said. "I warned everyone to think of everything that could go wrong, but no one suggested a hurricane!"

It was one of the worst storms of the century, and thus began several days of excruciating uncertainty. It had been hard enough to schedule the event so that Spielberg, Sagan and the equipment could be together at the same place and the same time. Now it looked like all our plans could be shot down and we would have to start all over again. But winter would be moving in, bringing with it fierce and unpredictable New England snowstorms.

I assured everyone that either (a) E.T. would keep the hurricane away, or (b) the eye of the hurricane would center on our radio-telescope, and then there would be an unearthly rumble and the Mother Ship would descend from space.

As I flew to Massachusetts, the hurricane moved in, raked the state, knocked out power in many towns, brushed the town of Harvard and departed. To our relief, the equipment suffered no major damage.

The day of the event was beautiful. It was a clear, fresh-smelling New England day, with the leaves of the lush countryside around the telescope just beginning to turn to their fall colors.

Spielberg arrived in a limousine with actress Amy Irving, their four-month-old son Max, and Carl Sagan and his wife Ann Druyan, together with their little daughter, Sasha. Sagan had held the infant Sasha in his arms when Project Sentinel was turned on two and a half years earlier, and now she was a dignified little girl growing up in step with SETI.

While the press was kept at bay, we took Spielberg and company through the rear entrance into the building housing the META electronics. In the control room, Spielberg's first reaction was "Hey, an oscilloscope!" As he watched the TV-like monitor with the waveform of a circuit signal on it, he explained that his father was an electrical engineer who used to let him fall asleep in front of an oscilloscope. I suddenly understood Spielberg's obsession with the gadgets and high tech shown in many of his films. Horowitz explained the functions of the different units, showing off his shiny new equipment with the pride of a new father, the near-sleepless nights forgotten.

Then we went outside to the big Frankenstein-type knife switch that Horowitz had dug up for dramatic effect. Amidst the crowd of reporters and cameramen, Spielberg, with little Max in his arms, approached the switch.

The turn-on of the META system at Harvard in 1985, with Carl Sagan (left) and Steven and Max Spielberg (right).

"It's such a nice day," said Sagan. "The telescope's not blown away. This is a moment in which the most sophisticated search for extraterrestrial intelligence in the history of the planet Earth is initiated." He went on to praise the remarkable technological skills of Horowitz, the generosity of Steven Spielberg, and the enthusiasm of the public—the product of whose efforts would be "an example of high-technology for solely benign purposes, and an example of what else would be possible."

Spielberg leaned over the switch and helped Max throw it, triggering the computer inside the building to load its program. A technician turned on the antenna drive. The great dish began to rumble and rotate slowly toward heaven.

"Well, I'm very happy to be involved in this project," said Spielberg, "because, as you all know, I've benefited so much from science fiction, I just thought it was time to get involved in science reality. And I'm real happy today to be here. And I guess everybody always considers space as a new frontier. I still think there are so many problems at home, that has to *remain* the frontier. But for Max's generation, when Max grows up, hopefully, he'll be the recipient of some outstanding information which the project will give. Something like, well, we're all the same.

"Beam us up here! I just hope that there is more floating around up there than just old reruns of the Jackie Gleason show." The antenna slowly moved, whirring.

A reporter asked Spielberg if he thought there was life out there.

"Yes, absolutely. The question is, is there life *here?* Especially, *intelligent* life?" Everyone laughed.

The exploration of the magic frequency of the universe had begun.

META Meeting

Following the turning-on of Project META, a symposium was held at Harvard, at which four scientists spoke: in addition to Paul Horowitz and meeting chairperson Carl Sagan, there were SETI founding-father Philip Morrison and physicist/science-fiction writer Robert Forward. META's quantum jump in SETI stimulated a wide-ranging discussion that gave a rare opportunity to witness the thinking of some of the most imaginative scientists in the world.

Sagan started the discussion by pointing out that the upgrading from Sentinel to META was really not all that expensive. It had cost about $100,000, which is big money for The Planetary Society, but small money compared to the kinds of technological projects that the U.S. Government routinely spends money on. One of the major consequences of this system would be to prod NASA, which has its still more sophisticated system underway, to do significantly better than META. And, to quote Sagan, "They will have to do *considerably* better in order to do better."

Paul Horowitz has the air of a boy genius about him. He'd be perfect as the young hero of a Spielberg movie who first picks up signs of another civilization. His enthusiasm is contagious and his impish sense of humor often surfaces unexpectedly.

He began by pointing out that there were two kinds of facts determining the future of META: hard facts and "facts about speculations." The first hard fact is that communication over galactic distances is possible. Basically, it's easy with the technology we have. He showed an equation describing information transmission: "The point I want to make is, this isn't speculation, this is communication-technology calculation. All I've actually done here says, if we had a couple of 600-foot dishes, well within our capability on Earth, and they were spaced a thousand light-years apart, could we transmit with a dollar's worth of energy a signal that could be received above the noise at the other end?

"At five letters per word, galactic telegrams cost a dollar per word. That's an amazing fact. It's a fact that interstellar telegrams are *cheap*. If you want to think of why, probably because the sky is very quiet at radio frequencies. Nature doesn't know how to make twenty-centimeter waves very well. And therefore, with modest-sized pieces of technology we can communicate."

So Horowitz then posed the question: what's within a thousand light-years of us? Within that distance, we have about a million stars like the Sun—that is, a million stars that could harbor life because we know the Sun does. Of course, there could be other conditions under which life can flourish. But we have to start with what we know.

Then he discussed the reason why he believes in searching at magic frequencies. "Here's the microwave region. Here's the place where the sky is quietest—therefore best for communication, the region from one to ten gigahertz. And right within that region are special spectral lines that can be guessed. Why transmit at a random frequency if you can transmit at a guessable one and make the job easier? We assume of course, that the beings transmitting to us are trying to make the job easy and not difficult. We're looking for a beacon."

The sensitivity of this experiment is such that one ten-billionth of a watt of radio power incident on the entire Earth would be detected by our antenna. It's the equivalent of a million Project Ozmas, and would detect our technology out to about a thousand light-years.

Horowitz went on to explain that the META required half a million solder joints, with 128 circuit boards that are each equivalent to a minicomputer. So powerful is the system that it is roughly equal to a Cray supercomputer.

"Although this is the most extensive search yet," he added, "it's not the ultimate, by any means. NASA in particular, with Stanford and Jet Propulsion Lab, plans a much more extensive search. But even a small fisherman can get lucky sometimes."

SETI pioneer Philip Morrison served as a mature counterpoint to the boyish Horowitz. Morrison, a former Manhattan Project physicist, has one of the widest-ranging minds in the world. Interested in all aspects of human culture, he speaks with a rapid-fire, staccato style infectious in its enthusiasm. In addition to his research in physics and astronomy, he somehow finds time to be the book reviewer of *Scientific American*, and is making a series for Public Broadcasting System television.

It was his enthusiasm that converted me from a physics major into an astrophysics graduate student, when our paths crossed during my last year as an undergraduate at M.I.T. and his first year as a professor there. He had just come from Cornell University, which is where I went the next year for my Ph.D. Prior to being exposed to his exciting descriptions of the wonders of the cosmos and the way that physics allows us to "X-ray" the universe, I'd always thought of astronomy as fun, but not a way for a physicist to earn a living. He changed my mind and my life.

He reminisced about the changes he has seen since he and Cocconi first suggested looking for extraterrestrial communications at the 1420 MHz line of hydrogen in 1959. Initially, he said, there was a small, enthusiastic SETI community. The public was generally skeptical at first, but then, gradually, more and more people became neutral . . . then favorable . . . then enthusiastic toward this remarkable endeavor. That growth is largely due to the progress of astronomy and the general growth of our interest in the stars around us. Much of this progress took place in the fifties and early sixties; after that, we began to sense that this was a project for which the time was ripe.

It isn't that great radiotelescopes have been built since that time. Of course, the Arecibo dish was not on the air then, but it was being designed. It was clearly ahead, and nobody has made a bigger dish yet. "Radio technology," Morrison points out, "is an *extremely* subtle proposition, and was well known. The enormous developments of the prewar and World War II years made it possible to do a rational design of the whole thing, and the limitation was really effort and money on a particular thing. Now, it can be done much more easily, much more cheaply. But in fact, the best receivers of today, the least noise, the highest figures of merit, are not a great deal better than they were in 1960."

What _is_ tremendously different is microelectronics. Morrison had seen instruments in nuclear physics labs that could analyze 24 channels at once, and as many as ten of these working together. They weren't by any means commonplace, but they were around. You could get perhaps a thousand channels at a time, and that seemed pretty good. But the most we might have foreseen, according to Morrison, would have been a hundred thousand channels.

Of course, here we are now—and through the rather modest, cooperative effort of a dozen people and some generous donors—tuning in to eight million channels, all at once. Certainly far more than we would have guessed twenty-five years ago. But what led to such astounding progress in so little time? According to Morrison, we can thank some very clever engineering ideas in the packaging of electronics. It was a typical development of high technology, engineering applied to the problem of putting electronics into smaller and smaller volumes. Not so much to make them portable and to have watches, which of course everyone does, but to make affordable, really powerful data processing in a box that is much bigger than a watch, but not any bigger than a refrigerator.

And, on top of these advances, we have reached the point with computers that we can now simulate a small part of the data-processing ability of living creatures. We can even simulate some of the higher functions of our own brains. "Not that rich, original, intuitive, complex, manyfolded thing that we think with," says Morrison, "but the kind of thing that you do when—and I hope I will offend no one in saying it—when you play chess, a limited thing for limited objectives, with a lot of intellectual power going to a small end. _That_ we can do really superbly, with a handful of chips or maybe more than a handful—a desk full of chips."

But what is so encouraging, Morrison maintains—though less important, by far, than the development of high-powered computer data-processing skills—is that "the so-called astronomical search that we talked about in early days now really looks like something we can accomplish without heroic efforts, without gigantic investments, only _considerable_ investments and _rather_ heroic efforts, and that's the kind of thing that we're going to go into now. Because it is still a big job, the biggest sorting and scrambling and pointing and arranging job that people have undertaken, and that's probably as it should be."

An Amazing Story

By sheer coincidence, the ceremony and symposium occurred on the same day as the premiere of Spielberg's TV series, *Amazing Stories*. Evidently, Spielberg's imagination was stimulated by his involvement with SETI. An episode of his series, dealing with a possible SETI discovery, was broadcast less then two months after he turned on META. Called *Fine Tuning*, it was from a story written by Spielberg, and showed a teenager picking up signals from another civilization.

It demonstrated an acquaintance with the ideas of SETI that most Hollywood science fiction does not: the first contact is by television, not spaceship; the civilization is ten light-years away, near a Sunlike star; and they receive the cream of our cultural broadcasts: *I Love Lucy, Burns and Allen, Milton Berle, Jackie Gleason, Bonanza*, and that pinnacle of western civilization, *The Three Stooges*. We know this because the teenager picks up alien versions of these shows.

"*We know all about your planet by intercepting television signals in outer space. We would have landed sooner, but we had to vaccinate ourselves against static cling, itchy scalp and ring around the collar.*" (© Cowles Syndicate, Inc. Distributed by King Features Syndicate, Inc.)

One of his friends summarizes the awe with which these transmissions are received: "I can't believe this! We're actually watching shows from outer space! Got any nachos?"

Drawn by this cornucopia of intellectual marvels, the aliens travel to Earth by spaceship at just less than the speed of light, so they arrive the day after the alien TV newscast depicting their takeoff. *Amazing Stories* gets an A for science in this one.

Fortunately, these tacky aliens are interested only in sightseeing in Hollywood.

And thus SETI, having profited by the success of *E.T.*, has in turn enriched the entertainment medium that nourished it. One additional irony now presents itself: this very show is now traveling through space at the speed of light. Perhaps one day, it will be picked up by aliens who may enjoy its message. Perhaps they will broadcast to us an alien version of *Amazing Stories* . . .

13

Beam Me Up, Mr. Spock!

INTERSTELLAR TRAVEL

What can be more palpably absurd and ridiculous than the prospect held out of locomotives traveling *twice as fast* as stage-coaches! We should as soon expect the people of Woolwich to suffer themselves to be fired off upon one of Congreve's ricochet rockets, as trust themselves to the mercy of such a machine going at such a rate . . . We trust that Parliament will, in all railways it may sanction, limit the speed to *eight or nine miles per hour*, which we entirely agree . . . is as great as can be ventured on with safety.
Quarterly Review, c.1825

SINCE WE HAVE four interstellar spacecraft right now, is human flight to the stars just around the corner? Is this how aliens communicate between civilizations, dropping by instead of phoning? Generations of science-fiction stories and films have practically convinced the average person that such travel is possible and that one day it will be as common as bus rides. Reality, however, is quite different—as best we can tell from the perspective of our incomplete knowledge of physics.

Einstein's theory of relativity says that nothing can be made to go faster than the speed of light, 186,000 miles per second. (There's a wonderful bumper-sticker that says: "186,000 Miles per Second Isn't Just a Good Idea—It's The Law!") If you build a super-powerful rocket, the theory of relativity says that the energy you put into the rocket will make it heavier and heavier the closer it gets to light speed, requiring more and more rocket fuel. It would take an infinite amount of fuel to reach the speed of light.

Having said that, I must point out that some scientists think there could be particles which were created by nature already moving faster than light. Thus they would not have to pass through the light-barrier, and may be permitted by relativity. Searches have been made for such particles, but they haven't been found.

Something just as strange has been found, however: parts of quasars, which seem to be exploding galaxies, have been measured as moving faster than light. This seemed at first to violate the laws of physics (and it upset a lot of physicists!), but it now appears that they are not really moving that fast; they are instead shooting out beams of radio waves that merely *seem* to move faster than light.

It is true that the nearest stars are a few light-years away, but I am confident we will go there eventually. The four spacecraft that are leaving the solar system will take many thousands of years to travel each light-year. At the very least, I expect that one day we will develop suspended animation to allow us to go to the stars even at such slow rates.

However, Einstein offered us a beautiful way out of this time-consuming difficulty, at least in theory. At the same time that he imposed a cosmic speed limit on us, he, like a crafty lawyer, included a loophole: the closer you travel to the speed of light, the slower time passes for you, an effect called time dilation. Astronauts, the fastest humans ever, return to the Earth a tiny fraction of a second younger than they would have been if they had remained home.

The scale of the solar neighborhood. From top to bottom, the scale increases by ten thousand times from each diagram to the next. At top, we see the distances of the Earth, Moon, and geosynchronous orbit (where communication satellites usually are placed). Next, the solar system. Finally, the nearest known stars: Proxima Centauri (orbiting around Alpha Centauri), Alpha Centauri itself (really two stars closely orbiting each other), and the next-nearest star, Barnard's Star. The Centauri stars are about four light-years away; Barnard's is about six. There may be even closer stars not yet found, such as the hypothetical dinosaur-killer, Nemesis.

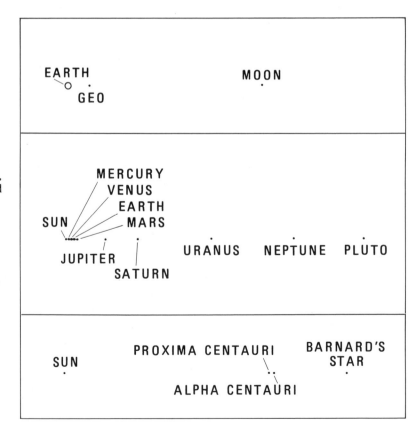

Some day, they may travel near the speed of light, in which case they might age only a couple of years for every decade they are away from Earth. This is an amazingly powerful effect when near the speed of light. So strong is it that, if you got in a rocketship traveling at a constant one-G acceleration—the same acceleration we experience on Earth and less than that occurring in some roller-coaster rides—you could tour the universe in a single human lifetime! This was the premise of a fine science-fiction novel, *Tau Zero*, by Poul Anderson. ·

The Energy Crisis

The META symposium debated the possibility of interstellar travel. Physicist Robert Forward was included in the meeting because his careful studies of possible methods for interstellar travel counterbalance the greater skepticism of most other scientists. Indeed, he is now advocating that NASA start planning its first probe to another star, an unmanned spacecraft he calls Starwisp.

Forward is a dynamic speaker with snow-white hair and a penchant for tuxedos and brightly colored vests. He is a physicist at the Hughes Aircraft Company Research Laboratories who has studied the measurement of gravitational radiation. He is also a successful science-fiction novelist, author of such books as *The Flight of the Dragonfly*, about an interstellar expedition using his spacecraft designs.

One technique discussed by Forward was first studied by Freeman Dyson and others years ago. Known as Project Orion, the idea was to power an interplanetary spaceship by throwing little atomic bombs out the back. The bombs explode and press against a pusher plate which pushes the spacecraft. A shock absorber smooths out the pulses so the astronauts do not experience dangerous accelerations. This turns out to be an effective way to travel around the solar system, taking much less time than conventional rockets. "It would have stored in there the world's supply of atom bombs," said Forward, "and I can't think of a better use for the results of a disarmament."

The project was dropped because it might violate international treaties. And, unfortunately, even our most powerful bombs can only get up to a *hundredth* of the speed of light. So even atom bombs aren't good enough.

What about fusion, the power source of the Sun and of hydrogen bombs? We are getting close to being able to harness this energy in the laboratory. But fusion machines, more powerful than fission bombs, aren't good enough to get us to the stars at a tenth of the speed of light.

Paul Horowitz expressed the view of most scientists when he discussed the most efficient rocket a physicist can imagine, one using matter and antimatter. Antimatter is the opposite of ordinary matter in every way. Every type of elementary particle has an antiparticle which is identical in mass but has the opposite charge. Thus, there are antiprotons, which are the same size as protons but are negative instead of positive. There are antielectrons (called positrons) which are like electrons, but positive instead of negative. Antineutrons also exist, though because the neutron has no charge, neither does its antiparticle. Antimatter is the mirror image of ordinary matter.

What makes antimatter of great interest for interstellar travel is that when a particle meets its antiparticle, they annihilate each other in a burst of energy. The energy comes from the energy that all matter has, according to Einstein's $E = mc^2$ equation. All matter has this enormous energy trapped within it, but it can be released only under special circumstances. Particle-antiparticle annihilation is the ideal circumstance: 100 percent of the energy is released.

For comparison, an atomic bomb releases only around *one tenth of one percent* of the energy contained in the uranium or plutonium atoms. (Hydrogen bombs are roughly ten times more efficient than atomic ones.) The bomb that destroyed Hiroshima caused the conversion of about one gram of matter into energy. That's about the amount of matter in a thimbleful of salt.

Antimatter is clearly the ideal fuel, in the sense that it contains the greatest amount of energy for a given amount of matter. Surely that is the best possible fuel for interstellar travel.

Currently, production of antimatter in physicists' particle accelerators is microscopic. But let's suppose you could produce antimatter in large quantities. And let's further suppose that you could contain antimatter somehow. You can't put it in an ordinary bottle, because it would annihilate the bottle, but you might be able to make a container using electric and magnetic fields so that the antimatter never touches ordinary matter except when needed.

Horowitz described the problems facing us if we want to send astronauts on a round trip to a nearby star such as Alpha Centauri. "Let's say we go at seven-tenths the speed of light so we don't age too much, using perfect matter/anti-matter engines of zero mass, zero overhead and so on—we can easily calculate how much energy that trip takes. It would cost 200 million billion dollars' worth of fuel at a cent per kilowatt-hour. It would equal the U.S. power consumption for a million years, and would be enough energy to power a 1000 megawatt radio beacon for half a *billion* years, which would be *my* preference, I guess."

And there are many other practical problems, including shielding the astronauts from the enormous amount of radiation that would be generated.

Forward was more optimistic about antimatter rockets. He has done a study for the Air Force on such rockets, and concludes that they are feasible. Also, antimatter is now being produced and stored (in minute quantities) in France. A similar facility is being planned for the United States. "And," said Forward, "I've also looked at the problems of turning antimatter into anti-hydrogen and storing it. And they are tractable. It should be possible to do it. Antimatter rockets *can* be built. But even antimatter rockets won't get us to the stars. I've actually gone through a rough design of an antimatter rocket, and you'd need—to get a ton of payload rendezvoused at a far star system—about 180 kilograms [400 lbs.] of antimatter. Which is not as much as you might think, but still 180 kilograms will take us an awful long time to make." Interstellar rockets seem to be enormously expensive and difficult to make, although perhaps a thousand years from now, this will not be the case.

Interstellar Gas Stations

There is a different approach that does not use rockets as such. The problem with rockets is that they must bring along their own fuel. This means that a rocket must not only have fuel to take the astronauts and their life-support equipment, but it must also carry fuel to accelerate the fuel. This is why the Space Shuttle is so big. It has to start with about 2000 tons of fuel to put 100 tons of Shuttle into orbit.

So what we want to do to go to the stars is to stop thinking about rockets. As Forward says, "Rockets are only one way of traveling through space. A rocket has a payload, some sort of guidance, a structure, an engine, some reaction mass that is used in the engine and some propulsion energy to heat that reaction mass up and throw it out the rear. Now, one of the things that I've been doing for the past ten years is trying to dream up rockets that don't have all these components, each one of which *weighs*. In looking for ideas of getting rid of the reaction mass, getting rid of the propulsion energy, the engine, the structure, the guidance, anything but the payload. In fact, Carl's latest book [*Contact*] has an example of an interstellar probe that has no reaction mass, no propulsion energy, no engine, no structure, no guidance—just payload. It happens to be a message

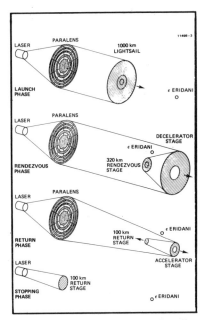

A three-stage interstellar lightsail, heading for the nearby star, Epsilon Eridani. The third stage (the center of the second stage) comes apart and is accelerated back toward the Sun by light reflected off the second stage.

which we receive and turn *into* a space vehicle by which we go back. So SETI in itself is rocketless rocketry, from my engineering point of view."

One such concept was suggested in 1960 by engineer Robert Bussard. Known as the Bussard interstellar ramjet, it would take advantage of the fact that space is full of hydrogen, fuel for fusion power. You would have to make a huge scoop, on the order of 600 miles in diameter, to sweep up the dilute hydrogen. This would be difficult to do, according to Forward. "If you can make a scoop out of 'unobtainium'—and that's unfortunately what it is right now—and weighing nothing, and yes, very high strength, and scoop up all this hydrogen, and put it into a fusion engine, without slowing it down first, because that would cause drag, and in this fusion engine create proton-proton fusion, which we don't have any idea of how to do now—the Sun does it, but we can't. If you could do all of those magical things, then here is a vehicle that would pick up its fuel as it went along and it would be able to get up to 99.99 percent of the speed of light, and you could travel through the galaxy in your lifetime, and come back, unfortunately, tens of thousands of years later. But still there's an *idea*. It is *not* a practical idea, and probably never will be."

Is there some more practical way to leave the fuel behind? Yes. The secret is to "beam the energy up." Then the spacecraft can use that energy to accelerate, and not have to carry fuel. Some research has been done that suggests it may be possible to lift vehicles off the Earth and push them into orbit using laser or microwave power. Robert Forward has done a great deal of research to see whether the principle can be applied to interstellar travel, and it looks promising.

Even Forward does not propose building interstellar spacecraft for astronauts yet. "But," he points out, "we can start thinking about sending small probes, not humans. We don't want to send humans—that's too costly and too expensive, and too dangerous. We have to spend a lot of money supplying them with water and entertainment and everything else that robots don't care about."

There is a useful number that puts the problem into perspective: if you were to travel for one year at one G—the acceleration of gravity—you would almost reach the speed of light. (For comparison, our *Pioneer* and *Voyager* interstellar spacecraft travel at the order of a hundredth of a percent of the speed of light.) However, Forward says that we don't want to travel at one G. That would waste a lot of energy. At one G acceleration, after you have been traveling for one year, all you're doing is adding mass to your vehicle (because of Einstein's relativity), making it harder to propel. So you don't want to keep accelerating. You want to get up to seven-, eight-, even nine-tenths the speed of light and *coast*. Doing this, you don't lose that much in travel time, and you save energy.

Forward's most practical design yet is the ingenious *Starwisp*, which could be humanity's first interstellar robot probe capable of returning data from a nearby star. His idea is to assume that a few decades from now, we have a microwave-powered solar satellite. It collects sunlight, turns it into microwave power and beams it down to Earth for power transmission. What Forward wants to do is to borrow that microwave transmitter for a week, during the testing phase before it's actually in use, and turn the beam the other way, running it through a special kind of a lightweight "lens" called a Fresnel zone plate.

This very large "lens" would focus the microwaves out at a distance and *push* a lightweight sail called *Starwisp*. This would be a mesh of wire, about a kilometer in size, weighing sixteen grams—less than an ounce. In each intersection would be a microcircuit. There would be a hundred billion of these microcircuits, not an outrageous number when you realize there may be ten thousand transistors in your watch at this moment. The total mass of the microcircuits would be four grams.

Forward studied the case of a 10-billion watt solar-power satellite, equal to two Hoover dams. This would accelerate a 20-gram spacecraft at 115 Gs—very fast. It would reach two-tenths the speed of light in a week and would arrive at Alpha Centauri in twenty-one years.

"Now," says Forward, "you've got twenty grams of wire mesh with microcircuits going through. What's it going to do for you? If you

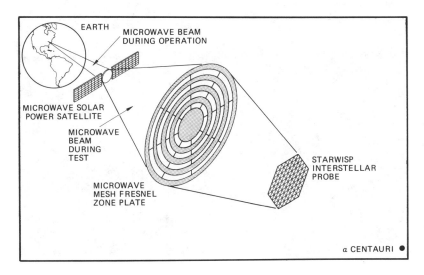

Starwisp: An unmanned interstellar space probe we may be able to build in the near future. It would use a proposed solar-power satellite (designed to beam power to Earth). For a brief time, the satellite would beam microwave energy to a very light-weight sail that would use this energy to accelerate this probe just like the lightsail.

design those circuits so that they use the wires as microwave antennas—they divide the wire structure up—they will collect power. I want to borrow the thing again for about thirty hours, twenty-one years later, and flood the Alpha Centauri system with microwave power. This mesh will pick up somewhere around ten watts of microwave power beamed from the Earth, so it doesn't have to carry its own transmission power.''

The microwave circuits would measure the phase of the microwaves as they went by and they would know where the Earth was. No pointing and tracking would be needed. Each one would have a diode that would look in a different direction at a different color with a fairly diffuse reflector. All of these circuits would take pictures of the planets and the star system as they went through, so we would get color television pictures back from this twenty

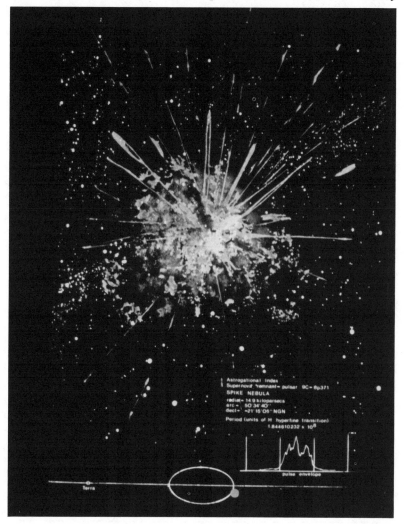

An imagined chart from an interstellar spaceship.

grams of circuitry—if one could make such a robotic probe. While Forward isn't certain it can actually be done, he wants people to know that there are reasonable ideas for making very small interstellar probes.

So human interstellar travel is possible, according to the known laws of physics. The history of technology shows that if something is possible, sooner or later it usually gets done. And usually much sooner than most people expect.

Rip Van Winkle in Space

I am convinced that, one way or another, we will personally visit other stars, if our civilization does not perish. It may be through the slow growth of colonization and exploration as we build worlds farther and farther from the Sun. Eventually we may industrialize the comets that stretch almost halfway to the nearest star. Billions of these comets are just waiting for us, filled with water and the other ingredients of life. As Sagan said, "A kind Providence has *filled* the Galaxy with refueling stops." ("Texaco stations," Forward calls them.)

It may be through the multigeneration starship beloved of science fiction. As an outgrowth of space colonization, people will spend their whole lives in space, and some of them may decide to take a self-sustaining space colony out of the solar system. They could wander through interstellar space, generation after generation, living and dying, until their descendants reached another star.

Or it may be through the development of human hibernation. There have been experiments in cryonics to find ways to freeze people in the hope that one day they may be revived. A few people have actually been frozen after they died, and are being stored in the unlikely event that future technology can not only defrost them, but reverse the damage done by the freezing process, cure whatever killed them, and bring them back to life. The main problem with this is that ice expands, so when bodies are frozen, ice crystals form in the cells, which expand and destroy them. But it may be possible to develop an antifreeze, or to use high pressures to prevent the destructive crystals from forming. One day, hibernation could become available, and then adventurers or colonists may be put into slow interstellar ships, to sleep away the centuries across the long light-years, and awaken in some brave new world.

Or we may travel to other stars on fast vessels, using Forward's principles or some as-yet unknown law of physics. One technique that no one else—not even Forward—has yet explored even

"Aaaaaaa! . . . No, Zooky! Grok et bok! . . . Shoosh! Shoosh! . . ."

theoretically, and which I believe may solve some of the remaining problems, is high-G acceleration.

Humans should be able to withstand much higher accelerations than we now experience in rocket launches. For example, immersing an astronaut completely in water, like a scuba diver, should cancel out most of the effects of acceleration. It may be possible to endure accelerations of hundreds of Gs for periods of weeks. For extreme accelerations, it may be useful to breathe water like fish, which humans have already done (using water to which extra oxygen has been added).

At ten Gs, you reach half the speed of light in a couple of weeks. At a hundred Gs, it takes only a couple of days. This technique would cut down the travel time. The high Gs need only be tolerated for the acceleration and deceleration periods. Most of the trip would be spent in zero gravity, or if more convenient, in artificial gravity produced by spinning the spaceship. Furthermore, since unmanned payloads can be accelerated much faster, interstellar astronauts could be resupplied from Earth.

Traveling to the stars will be difficult, time-consuming and expensive. And it may never be the fast, routine affair of science fiction. But some day, I believe, daring humans will look outside their spaceship and their excited faces will be lit by the glow of the three stars of Alpha Centauri.

Are They Here?

UFO'S AND OTHER EVIDENCE

Kids get their ideas about UFO's where I learned about sex: the tabloids and the sleazy press.
J. Allen Hynek

ON THE AFTERNOON of June 5, 1969, two airliners and a National Guard fighter near St. Louis encountered a squadron of *UFO's*—Unidentified Flying Objects. They looked like hydroplanes of burnished aluminum. A Federal Aviation Administration official in the cockpit of one of the airliners said that it seemed as if the UFO's were about to hit the plane, and that they were only a few hundred feet away. St. Louis radar confirmed the UFO's.

The fighter was nearly hit, radioed its pilot. He reported that the UFO's seemed to take evasive action, as if they were "under intelligent control."

Sounds like a scene from *Close Encounters of the Third Kind.* But this was reality, folks. This was one of those chilling reports that has given rise to hundreds of books and thousands of newspaper articles since the first "flying saucer" report.

However, the St. Louis UFO quickly became an *IFO* (Identified Flying Object), thanks to a photographer, Alan Harkrader, who took a picture of it, and to Philip Klass, senior editor for *Aviation Week and Space Technology*—the Sherlock Holmes of UFO's.

Author of *UFO's Explained* and *UFO's: The Public Deceived*, Klass proved from Harkrader's photographs, many eyewitness reports and interviews with the aircraft personnel, that what they had all seen was indeed from outer space—it was nothing but a meteor! His investigation showed how even experienced observers can be misled by unexpected phenomena.

"Yeeeeeeeeha!"

183

He also proved that the meteor was at least 125 miles away from the aircraft. Many people find it hard to believe that experienced pilots could make such a mistake, but they fail to realize that (a) it is extremely rare for anyone to see a bright, daytime meteor; (b) no one, no matter how experienced, can judge the distance of a faraway, unfamiliar object with the naked eye. A strange, glowing fireball such as this meteor could have been a tiny object nearby or a huge object far away. In reality, the glow from even a small meteor can be bright enough and big enough to be seen hundreds of miles away.

What about the radar detection of the St. Louis UFO? Klass found out that, at that time, the airport did not yet have radar that could measure altitudes. Furthermore, they did not mark radar blips except when a plane radioed that it was coming in to the airport. Since none of the three planes involved was going to land at St. Louis, their blips were unmarked. When one of the planes radioed the UFO report, he looked at his screen, identified one blip as the plane that had just called, and sure enough, there were two unidentified flying objects! He didn't realize they were the other two planes that were watching the UFO's.

Such anticlimactic outcome has usually been the case whenever a UFO report is investigated thoroughly. Despite this, a 1986 survey by the Public Opinion Laboratory at Northern Illinois University found that 43 percent of the public agreed with the statement that "it is likely that some of the unidentified flying objects that have been reported are really space vehicles from other civilizations." This seems to be wishful thinking.

Nevertheless, it is legitimate to ask whether every UFO report can be explained away.

To most SETI scientists, UFO's are a can of worms into which they do not want to stick their hands. So filled with fraud, misidentification and pure lunacy is the UFO business, and so time-consuming is a proper investigation, that most scientists avoid it. SETI researchers already have the difficulty of often being confused in the public mind with the nuts who think we are being visited by little green men from Venus.

But it would be improper, in my judgment, to discuss SETI without at least touching on this subject. The single most common question I am asked by nonscientists when the subject of SETI comes up is this: "Do you believe in UFO's?" This chapter is my answer.

The first part of the answer is this: yes, I "believe" in UFO's—*unidentified* flying objects. I have often seen UFO's, and so has the reader—objects in the sky which, for a moment, look strange, but

"They don't allow those on my planet."

which turn out to be airplanes, birds, hot-air balloons, meteors, strange clouds, Venus, the Goodyear blimp, and so forth. They were UFO's for a moment, until they became IFO's. Every one of these IFO's in my list has generated many UFO reports around the world, which investigators were eventually able to identify.

A 1973 Gallup poll indicated that 11 percent of the American public has seen a UFO, but you may be confident that most of these could be easily explained away, since most people are not trained observers and have no idea of the incredible variety of atmospheric, aeronautical and astronomical phenomena that may be seen in the sky once in a blue moon. Under unusual conditions, mundane phenomena can fool even trained observers. Even the most optimistic UFO investigators have usually concluded that 80 to 90 percent or more of UFO reports are just misidentified normal phenomena.

I have never personally seen any UFO that did not turn quickly into an IFO. But I am still willing to admit that other people may have seen something heretofore unknown. And given that many scientists think that life may exist elsewhere, and that interstellar travel is at least conceivable, the possibility of extraterrestrial visitors should be considered.

Close Encounters of Many Kinds

Strictly speaking, there have been UFO reports for thousands of years, going back at least to the bizarre reports of an extraterrestrial-like visitor in the Bible's Book of Ezekiel. But it was an incident on June 24, 1947 that threw *flying saucer* into our language and inspired the modern fascination with UFO's.

A private pilot named Kenneth Arnold was flying near Mt. Rainier, in the state of Washington. Suddenly, he saw nine circular objects flying rapidly in a diagonal formation, weaving around the mountaintops. He later described this to a reporter by saying they "flew like a saucer would if you skipped it across the water." The story was carried by wire services all over the country, and the term *flying saucer* was born.

The war had ended just two years earlier, with its atom bombs, long-range missiles and other pieces of science-fiction technology-turned-reality. The Rand Corporation, just a year before, had reported that it would be possible to send rockets into orbit around the Earth. The world was ripe for extraterrestrial visitations.

In the weeks following Arnold's report, flying saucers were reported in every state, and in Canada, England, Australia and Iran.

Since then, thousands of UFO reports have emerged from virtually every country in the world, and many of them have been investigated by scientists and journalists. Many have wanted to prove that we are being visited by creatures from another world; some have wanted to prove that we are not. Philip Klass is probably the most diligent of the skeptics. He has found mundane explanations for numerous UFO's that less thorough investigators concluded were extraterrestrial spacecraft. He believes that every UFO case could be explained away if investigated with similar care. His attitude is common in the scientific community, although a survey by astrophysicist Peter Sturrock showed that the majority of the members of the American Astronomical Society thought UFO's were worth scientific study. (Twenty-one percent thought they were probably unworthy of such research.)

There was a respectable scientist who often locked horns with Klass on the UFO issue. His name was Allen Hynek, and he was a reputable astronomer who devoted much of his time to investigating UFO's. Initially, he was a skeptic who worked on the now-defunct Air Force Project Bluebook, studying UFO reports. But eventually, he became convinced that some UFO reports may not be explainable by known science. In the interests of a balanced viewpoint, I interviewed him in 1984, two years before his death, after his retirement from Northwestern University.

Hynek was also the technical advisor for Spielberg's *Close Encounters of the Third Kind*, and even appears as one of the scientists at the climax of the film. A vigorous man in his seventies, with a white beard, he looked the part of a distinguished scientist. I was delighted to discover that he had a noisy parrot who whistled the theme from that movie. While the parrot squawked, we talked of the fear SETI scientists have of being mixed up with UFO's.

"I certainly hope you get results," he said. "You see, the paradox, dilemma, in this whole UFO/SETI business is that, as an astronomer, I think it would be preposterous to think that we are the only intelligence. It's ridiculous! But that does not mean at all that that's synonymous with the UFO phenomenon, and this is a hard point to get across. I don't talk about UFO's very much anymore, I talk about the UFO *phenomenon*. The phenomenon is something that even Phil Klass can't deny. The phenomenon is the continual flow of reports, now from 140 countries—and our databank now has over 100,000 entries—and the patterns in those reports.

"The patterns are the intriguing things, because you don't get, for instance, reports with UFO's with wheels. And then there's the spectrum of witnesses—some of them remarkably high in technical background. We're especially interested in the cases where there

"There's four billion of them down here! Make sure none get on the ship before we take off."

are independent witnesses when there's no possibility of collusion—witnesses who don't even *know* each other, yet they're reporting essentially the same thing. Now, *that's* the phenomenon."

Hynek's critics often accused him of believing that UFO's are spaceships from another civilization. Ironically, he was himself a strong critic of interstellar travel. Evaluating the UFO phenomenon, he said, "I haven't the slightest idea what's behind it. It may be some aspect of our own psyche and intelligence. We might have to look for some possible interface between our physical dimension and another."

People often say we've gone to the Moon, so why can't they come here? "Well, it's ridiculous, as far as our technology is concerned. My standard ploy in that is I'll take an ordinary playing card and I say, now let the thickness of this card represent the distance from Earth to the Moon, a distance which we've actually gone. How many cards would you have to put back to back to get to Alpha Centauri? Nineteen miles of them! There ain't no *way* of getting here from there. But this is what makes people ridicule or downgrade the UFO phenomenon.

"I'm interested in studying the properties of the phenomenon as much as a mathematician might be interested in studying properties of a function. The demographics of it—what sorts of people report, what are the patterns in it?"

J. Allen Hynek, UFO investigator, examining a rock sample from a reported UFO landing site.

Hynek liked to draw on his interest in the history of science for perspective. Any civilization, he suggested, will interpret any new phenomenon in terms of its own technical development. Therefore, it's only logical that the public today would say "it must be somebody else's spacecraft." However, it might be something totally different. Hynek pointed out that a hundred years ago we had a perfectly well-known phenomenon—namely, the Sun. But nobody then could understand why, because nobody knew that the atom even had a nucleus. So perhaps we're facing a similar unknown. We have a strange phenomenon—there's no disgrace that it can't be explained. But the disgrace, as Hynek put it, is to put the cart before the horse and make UFO synonymous with spacecraft from elsewhere. "The public in general thinks that Hollywood has solved the problem. UFO's are spacecraft piloted by E.T. Fine, let's go and play bridge now or something else. The problem's solved. Well, it's *not* solved."

It was difficult for Hynek to get scientists to take his work seriously. "All it takes is one idiot to say he's gone to Venus in a flying saucer and have that published in the *National Enquirer*, and it sets back serious work on UFO's tremendously."

I was curious what had turned this man from an ordinary astronomer into a UFO investigator. He taught the history of science at Ohio State for a number of years, and it always impressed him how it's the things that "don't fit" that lead to the breakthroughs. Things that fit lead to another decimal place. It's the paradoxes that fascinate Hynek—hence his great respect for SETI and his sincere hope that serious UFO research and SETI research will not be enemies. After all, he points out, we're both searching.

So, in the best of all possible worlds, how would he go about investigating UFO's? Hynek said that to properly investigate a case would take about as much money as to investigate a homicide. It would require laboratory tests, interrogations, travel and controlled samples, for instance. You have to have hard cash to get hard evidence. You've got to go out and do the job properly. "What has happened is, of course, we have a bunch of amateurs, well-meaning amateurs, who spend weekends, just 'dilettanting.' But they haven't the least idea how to—they don't understand the scientific method. They will lead witnesses, all that sort of thing." This is just another reason why scientists are so skeptical about many reports.

What would Hynek have done if unlimited money were available? The first would be to hire trained statistical workers, and probably psychologists as well, to examine the enormous amount of data. That's what he calls the "passive phase."

But what intrigued him most was the "active phase." There, if money were no object, he would have hired and trained about a dozen good people who have training in both the physical sciences and the soft sciences. Out of the welter of cases he sees each year, he would select five or six. He would then assign two people to each of those critical cases, and it would be their job to bird-dog that case if it took two days, two weeks, two months or two years. It would be their profession. And they would write a technical report that would be of the caliber that could be presented to the National Academy of Sciences.

"If we had one-tenth the funds that the FBI spent in tracking down Patty Hearst, for instance, we could get something," said Hynek. "But there's never been a case with what I call a real FBI-type treatment, because it takes funds. That's what'd be needed."

Finally, I asked him if his research had inspired any ideas about SETI.

"Of course, way in the background is that gnawing feeling that we've had radio just a short while and radio communication might be to them what smoke signals are to us."

I remarked that I sometimes imagine there's a little extraterrestrial club somewhere, of beings who like to use primitive technology, the way some modern humans use spinning wheels as a hobby. The hobby of these extraterrestrials would be communicating with primitive civilizations, and they might transmit signals as eagerly as our ham radio operators do, in the hopes that one day they will make the first contact with a new civilization. Different worlds might even compete to be the first.

Hynek said, "I hadn't thought of that. But I like that idea. That's a possibility."

Before we parted, he added this message to his terrestrial colleagues: "In science, there should always be room for honest disagreement. I would much rather talk with an intelligent skeptic than with a dyed-in-the-wool believer who will accept anything and buy the Brooklyn Bridge tomorrow. Pseudoscience is a very real danger in our society, no doubt about it. But we should look at the possibilities. The playing field of science should have some bleachers, where we can have some of these things on the fringes, not to be ridiculed, but to be looked at. To be looked at."

What to Do with UFO's?

I must add a personal footnote to the UFO debate. While it is absolutely essential not to be taken in by the UFO phenomenon, the history of science is filled with phenomena that were ignored by scientists and other intelligent people simply because they did not yet fit into the current view of the universe.

Meteors, for example, were thought for centuries to be an insubstantial atmospheric phenomenon like lightning. Several well-documented observations were uncovered during those centuries which showed that meteors had hit the ground, leaving craters and pieces of rock. Yet noted scientists—and even Thomas Jefferson—ignored the evidence that rocks do fall from the sky.

Therefore, it seems to me that we should adopt an objective attitude toward UFO's, investigating them as we would any other unknown phenomenon. We may find that there is nothing there. We may find interesting psychological phenomena at the basis of some sightings. But perhaps we may find some real phenomenon, not necessarily beings from another world, but perhaps some piece of new information about our universe that is worth having.

But let us never forget the perils involved. Klass reports a highly intelligent observer who saw a UFO twice. The object flew around the night sky, blinking. The second time, with his son, he filmed it with a movie camera he had brought along in case the phenomenon repeated. "I could hardly contain my emotions," he wrote to Klass.

A friend of his saw it, too. A beautiful report: three observers, motion picture evidence. The friend, however, chased the UFO, and got close enough to see it clearly. She found a sign on it, not in an alien language, which said, "Anthony's Auto Body—Free Estimates." It was a small plane with strings of electric lights spelling out the message.

So the last part of my answer to the question "Do you believe in UFO's?" is this: I don't know of any absolutely reliable evidence that we have been visited by beings from another world. Most famous UFO cases can be explained by mundane phenomena, and the few remaining ones are filled with uncertainties and lack physical evidence.

UFO's should be investigated, and I applaud the courage of people like Hynek who risk the scorn of their colleagues. But the investigation must be done with especially great care. It is too easy to be fooled—by Mother Nature or by a fellow Earthling.

15

Where Have All the E.T.'s Gone?

SCIENTISTS AGAINST SETI

The absence of evidence is not evidence of absence.
Martin J. Rees

FOR DECADES, it was difficult to get the scientific community to take the idea of extraterrestrial intelligence seriously. Then, following *Sputnik*, pioneers such as Cocconi, Morrison, Drake and Sagan put forward arguments so persuasive that the tide of opinion gradually shifted to the point where many scientists felt the concept was at least plausible, if highly uncertain. This was a revolution in attitudes, and it was followed by a counterrevolution. There is now a scientific movement trying to convince us that life is rare and that we may, after all, be alone in the universe. If they are right, SETI is a waste of time and money.

The first major battle in the modern counterrevolution was initiated by scientist Michael H. Hart. In 1975, he published a paper entitled *An Explanation for the Absence of Extraterrestrials on Earth*, in which he started out with the premise: *"There are no intelligent beings from outer space on Earth now.* (There may have been visitors in the past, but none of them has remained to settle or colonize here.)" He contended that if there were extraterrestrial civilizations, then during the billions of years they have had to circulate, they would have been here by now.

This was actually a revival of an idea of the great Italian physicist Enrico Fermi. In 1950, at Los Alamos Scientific Laboratory, he had asked his colleagues, "Where are they?" (The Hungarian physicist Leo Szilard replied, "They are among us, but they call themselves Hungarians.") The argument has since been dubbed the Fermi Paradox.

Hart later investigated the range of distances from a star over which a planet can have a climate neither too hot nor too cold for life. He concluded that Earth was extremely lucky: had we been just slightly closer to the Sun, we would have been too hot, like Venus; slightly farther, and we would have been too cold, like Mars.

Scientists are theoretically supposed to be cold, dispassionate analysts of competing theories and experiments, much like Mr. Spock of *Star Trek*. But the reality is often quite different, and in 1980, a man came along who stirred up a hornet's nest of argument within the SETI community. Frank Tipler, then of the mathematics department of the University of California at Berkeley, wrote a paper whose title bluntly stated *Extraterrestrial Intelligent Beings Do Not Exist*. In this and several later papers, he extended Hart's arguments, repeated the claims of some biologists that the evolution of intelligent life is extremely difficult, and added two new twists, concluding with a statement guaranteed to provoke the SETI community: "The argument seems conclusive to me. I think we shall have to accept the fact that we are unique in the universe."

The first new twist that Tipler added to the anti-SETI argument was to insist on the *Anthropic Principle*, which Tipler phrased as this: "Many aspects of the Universe are determined by the requirement that intelligent life exists in it." This principle is an idea that some philosophers and scientists have toyed with over the centuries. Today, we can say that if the universe were much younger, we would not be here. If the laws of physics were slightly different, carbon would not be abundant and our life could not exist. This much does not raise eyebrows among scientists.

However, Tipler goes further and cites a calculation by the physicist John Wheeler that if the universe had been much smaller than it is, it would have collapsed by now. To Tipler, this is evidence that the universe is just barely large enough for us, and he asserts that "The universe must contain [as many stars as it has] in order to contain a single intelligent species." He thinks the universe seems designed specifically for human life. Tipler might paraphrase Descartes' "I think, therefore I am" as "I think, therefore the universe is."

Tipler's second new wrinkle was inspired by one of the founders of computer science, John von Neumann. In the early days of computers, von Neumann speculated about the possibility of machines that duplicate themselves, or self-replicating machines. Even then, it was clear to him that eventually a computer would be made that could make other computers.

Tipler suggested this: sooner or later, a civilization will arise that builds such self-replicating machines. It develops space travel and sends its self-replicating machines to the stars, along with robots that can mine planets and process materials. The machines travel to the nearest stars, dig up worlds, duplicate themselves and send the duplicates on to the next nearest stars. The duplicates do the same, and before long, just as Adam and Eve were supposed to have begotten the billions of people alive today, billions of von Neumann machines would be digging up every useful planet. They would be the ultimate real-estate developers.

Even if they traveled at considerably less than the speed of light, they could redevelop the entire galaxy within the age of the Earth. The fact that we were apparently not redeveloped, and could not even see any signs of their presence, Tipler argued, proved that we were the first civilization.

The pages of scientific journals were afire with replies to Tipler. How do we know that extraterrestrial civilizations would *want* to build enormously complex, expensive machines whose sole purpose is to mindlessly build more machines? Perhaps advanced civilizations would have advanced ethics. Maybe they don't permit interference with developing civilizations; maybe we are "quarantined" (a concept from science fiction: the Zoo Hypothesis). We could be living in a galactic zoo and not know it. And perhaps building interstellar von Neumann machines is not as easy as Tipler thinks. Or maybe if such a sick civilization starts destroying the neighborhood, its more civilized neighbors put a stop to it.

Some of the many possible answers to Hart's and Tipler's arguments are explored below.

The Planet of the Thermostat

Hart's argument about the narrow range of survivability of the Earth touched on a mystery of our history that we are now beginning to unravel.

One of the puzzles of the origin of life on Earth is how the climate was warm enough for life way back then. The earliest lifeforms seem to have been adapted to hot water. And geologists who have been able to measure the temperature of ancient Earth by studying the behavior of radioactivity in rocks find that the Earth was, if anything, warmer than today.

So what's the problem? The problem is that the Sun should have been much cooler back then. We can estimate its brightness because we think we understand how the Sun works. It uses the same type

One of the most unlikely telescopes on Earth: the neutrino telescope of Raymond Davis, Jr. (University of Pennsylvania), deep inside the Homestake Mine, South Dakota. The large vessel contains 100,000 gallons of a chlorine compound. Once in a great while, a neutrino from the Sun hits a chlorine nucleus, converting it into argon, a gas detectable by its radioactivity. This was the first experiment to tell us that something is seriously wrong either with our theory of the Sun or with our understanding of neutrinos—we still don't know which.

of mechanism as the hydrogen bomb: protons (the nuclei of hydrogen atoms) slam into one another at temperatures of millions of degrees, eventually forming helium. The transformation of hydrogen into helium generates the Sun's energy.

But helium is denser than hydrogen, so as hydrogen is converted to helium, the Sun shrinks. In shrinking, the compression of the gases by gravity heats the Sun up, so it gets brighter. Thus, the Sun must have been dimmer in the past—about 30 percent dimmer when the Earth was formed. That is so dim that the oceans should have frozen over, and life as we know it could never have formed. This problem has earned the nickname The Faint Young Sun Paradox. Paradox.

What's wrong? One possibility is that we don't really know the Sun as well as we think we do. One of the most unusual observatories in the world suggests this may be the case.

This observatory is in the last place you would think to put a solar observatory: at the bottom of a mine shaft, miles underground. The reason for this location is that astronomers want to see a byproduct of the Sun's nuclear reactions, neutrinos. Neutrinos are particles far smaller than even the tiny electron. They have given scientists numerous headaches—and a Nobel prize—because they are so hard

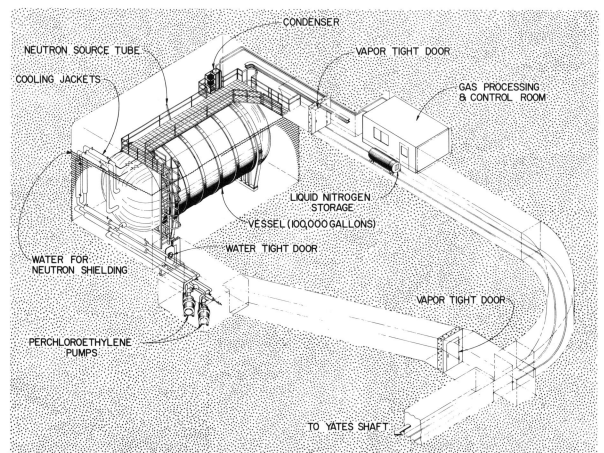

The neutrino telescope of Raymond Davis, Jr.

to detect. They have no electrical charge or magnetic field, so they don't interact with most other particles. The only way to detect them is to build a huge device with enough atoms in it so that the occasional neutrino that does collide with a nucleus can be found. The detector must be shielded from cosmic rays (other energetic particles from space), and the best place to do that is at the bottom of a mine. Raymond Davis, Jr., of the University of Pennsylvania built such a detector with the help of Brookhaven National Laboratory, in the Homestake Mine of Lead, South Dakota.

There, where a mile of Earth absorbs most interfering cosmic rays, he eventually found solar neutrinos. Unfortunately, he found only about a third of the expected number. Other scientists with similar detectors have confirmed his discovery. Where are the missing two-thirds?

We don't know.

Either something is wrong in our understanding of neutrinos (perhaps they transform into other particles), or there is something wrong with our theory of the Sun. If the latter is true, it might be that the early Sun could have been warmer than we now think.

The other way out of the Faint Young Sun Paradox is the possibility that the Earth was warmer because of the so-called greenhouse effect. This is the idea that some gases can act as a blanket, keeping a planet warmer than it would otherwise be. Venus, for example, is very hot not only because it is closer to the Sun, but also because its atmosphere consists mainly of carbon dioxide. This gas is nearly transparent to sunlight, so the light passes through to the cloud-tops near the planet's surface. This intense light heats the clouds and atmosphere, which in turn heat the surface.

A planet's temperature is usually determined by an equilibrium between two processes: the sunlight coming in and the heat radiated away. The sunlight is absorbed, the surface gets hot, and hot objects radiate most of their energy as invisible infrared light. And it happens that carbon dioxide, though transparent to visible light, is opaque to infrared. The carbon dioxide absorbs the infrared.

Infrared photons struggle through the atmosphere, being absorbed and re-emitted, and eventually escape to space. But for a while, the infrared heat is trapped as if a blanket were tossed over a heat lamp. Hence Venus gets not just hot, but hellishly hot.

The same thing could have happened on Earth if there had been lots of carbon dioxide back in the early days. (In fact, it may be in our future as well as our past. If the world population keeps growing, and continues to burn wood and fossil fuels into carbon dioxide, the resultant greenhouse effect may warm the planet enough to melt the ice caps in the next century and flood all our coasts!)

Until recently, however, we did not think carbon dioxide was abundant on the early Earth. Ammonia, if abundant enough, can also create a greenhouse effect, but it has trouble existing in large enough abundance. Sunlight tends to break much of it down into other gases.

The Faint Young Sun Paradox has forced us to consider other conceivable early atmospheres, and just as Jupiter and Saturn suggest a possible methane-ammonia atmosphere, Mars and Venus, both having carbon dioxide atmospheres, suggest another possibility.

In fact, Venus is particularly intriguing. Once called Earth's twin, it is similar in size yet radically different in atmosphere today. Why is it so different?

Soviet and American space probes have landed on the surface of Venus and Russian-French balloons have floated in the air. They measured the air pressure and found that it is about ninety times that of the air we breathe on Earth. Ninety atmospheres of carbon dioxide! On Earth, that's the pressure in the ocean at a depth of more than half a mile! Where did all this Venusian carbon dioxide come from, and why don't we have it on Earth?

It turns out that we do have it on Earth, but it's hidden. It's in rocks. Shellfish and plankton long ago converted carbon dioxide into shells. Shells from the dead critters rained through the ocean, forming sediments which eventually became rocks such as limestone. If you took the carbon dioxide out of the rocks, Earth would have very nearly the same atmospheric pressure as Venus—and most of our air would be carbon dioxide.

If Earth managed to keep just a few tenths of an atmosphere of carbon dioxide during its early days, that would be enough to keep the average global temperature above freezing. This also suggests why Earth has had such a happy history: never too hot, from a runaway greenhouse effect; never too cold, from a big freeze. The reason may be that the whole planet acts as a giant thermostat.

When it gets too cold, the shellfish stop growing so well, and they don't take so much carbon dioxide from the air. Deep inside the Earth, carbon dioxide is continually produced by the breakdown of the shell-rocks. Thus production of the gas goes up, the greenhouse effect swings into action, and a balmy climate is still possible.

On the other hand, if things get out of hand and the carbon dioxide threatens to turn Earth into another Venus, the gas has the useful property that it scatters sunlight much more effectively than ordinary air. The same phenomenon that makes the sky blue—Rayleigh scattering, the deflection of light by molecules—causes more sunlight to be reflected from the Earth, making the temperature fall.

If this theory is right, here's what your local weatherman might be saying tomorrow: The Sun will continue to heat up, but average temperatures are expected to be moderate for the next few billion years. You won't need so much suntan lotion, thanks to the rise in the level of carbon dioxide. However, if you're planning to sightsee in New York or London or any other city near the shore, be sure to bring along your scuba gear.

This should be a reliable forecast until the Sun does something really weird, like turn into a red giant star when it runs out of fuel.

Answers to Fermi?

So at least part of the answer to the critics of SETI may be at hand: life itself may protect a planet from drastic changes that would kill off life. But what about the other arguments?

First consider Tipler's argument of the Anthropic Principle, which contends that, because we exist, the universe must be the age it is, and it must just be big enough for us—and only us—to come into existence.

This, it seems to me, is just an age-old fallacy dolled up in modern dress. Originally, the fallacy was that the Earth is the center of the universe. Up until Copernicus, almost everyone thought that the universe revolved around us. When Copernicus showed otherwise, most people resisted because it challenged the religious dogma that their lives revolved around, and because it was humiliating to think that we might *not* be the center of the universe.

Then, when Darwin showed that we were cousins of the apes, many people resisted again because the idea was so humbling.

Now Tipler would have us believe that the universe is large enough for only one intelligent species. Maybe that is true. But surely it is much too soon to conclude that. Surely it is not unreasonable to think that a universe with billions of galaxies, each of which contains hundreds of billions of stars, might have plenty of room for many creatures.

To jump to the conclusion that they are not there simply because we don't see them easily is to make the same mistake people made about microscopic life. Until Leeuwenhoek invented the microscope, it was thought that nothing smaller than an insect or a mite lived. The world was crawling with zillions of microscopic beasties, but because nobody had seen them, they did not exist. The same may be true of aliens.

Tipler's other major argument has von Neumann machines roaring through the galaxy, multiplying like mosquitos in a swamp. There are many possible explanations of why we don't see such signs, besides the possibility that we are alone. But first a word in his defense.

If space colonists or their robots move at, let us say, a thousandth of the speed of light, it would take them only 100 million years to spread from one end of the galaxy to the other, about 1 percent of the age of the galaxy. Thus, if such a civilization had arisen a few billion years ago, they ought to have been here by now. It is thus perfectly reasonable to ask Fermi's question: Where are they? And it is not far removed to ask Tipler's version of this question: Why don't we at least see their machines right now? The possible answers scientists have come up with lead to some of the most fascinating speculations in all of science. Here are a few of them.

One possibility is that they *have* been here, but didn't see any interesting tourist attractions a billion years ago, and so left.

Another possibility is that they seeded life, as Fred Hoyle, Francis Crick and others have suggested. We are then their descendants.

Yet another possibility was once suggested jokingly by Thomas Gold: Maybe they landed before life arose, dumped their trash, took off, and the contamination from the trash was the seed of life which later evolved into us. This is sometimes called the Garbage Hypothesis.

In addition, Carl Sagan and William Newman have calculated that for a reasonable population growth rate, the galaxy is too big. They're not here yet.

Or maybe they are here *right now*, but don't want to make themselves known. This hypothesis might explain the rare UFO report that is otherwise hard to understand.

A variant of this theory was suggested by Michael Papagiannis, the first head of the SETI Commission of the International Astronomical Union. He explored the possibility that very advanced civilization would outgrow the need for planets. Interstellar travelers might become so accustomed to living in space with its zero gravity and controlled environment that they might find outer space much more pleasant to live in than dirty planets with poisonous atmospheres, dangerous bacteria and viruses, and heavy gravity. They might only drop by planetary systems to replenish their supplies, in which case the best place to do that in our system would be the asteroid belt. There, they would find a supermarket of resources easily accessible: broken-up worlds that have no bothersome gravity to deal with. Papagiannis proposes to study the infrared data of the IRAS telescope to find unnaturally warm objects.

Anthropologist Ben Finney of the University of Hawaii has studied the history of human migrations and co-edited the book, *The Los Alamos Conference on Interstellar Migrations*. He offers some lessons that may answer Fermi and Tipler.

Hundreds of miles north of the Hawaiian islands, archaeologists have found the remains of Polynesian settlements that became extinct long before Europeans came. The Polynesians who first settled Pitcairn Island had nearly vanished by the time the *Bounty* sailors landed. And consider Easter Island, whose great, brooding stone heads were built by ancient Polynesians. After the island became overpopulated, they were decimated by war, ecological disaster and starvation. They never escaped to another island. "The moral of this Polynesian tale," Finney writes, ". . . might be that running off to distant new worlds resolves nothing. It only transfers the inevitable population/resources crunch to a new setting—one which may . . . leave the people without the desire or means to flee farther . . . A space migration future might be strewn with many such dead ends."

Ming Dynasty China is another thought-provoking case. For a short time, China became the leading sea power in the world, much like Great Britain. From 1405 to 1435, they sent huge ships (up to 500 feet long) across the Indian Ocean and to the east coast of Africa. Then, apparently something internal happened that stopped them. They forbade oversea trading, and made it a capital offense to build seagoing junks with more than two masts.

What caused this radical change? It may have been the victory of land-oriented bureaucrats over outward-looking Imperial eunuchs. It may have been that the building of canals caused a focusing of energies back within China. Or it may have been a revival of Confucianism, causing them to turn inward rather than outward. Perhaps philosophy is the ultimate barrier to relentless conquest. "Mere possession of the technology for expansion is not enough," says Finney. "The motivation to expand must also be there."

Others have suggested that interstellar colonization may take much longer than Tipler thinks. Not only are the stars far apart, but the real history of expansion would not be in simple, straight lines. A colonizing civilization would favor the nearest nice star, which might be off in some odd direction. This would be true at each stage of expansion, so interstellar colonization would be along zigzag lines. Plotting these more realistic paths shows that it would take far longer to colonize the galaxy than a straightforward calculation would suggest. "In that case," says Frank Drake, "if you work out the numbers, it turns out that the time required to cross the entire galaxy is just equal to the age of the galaxy. And what that means is that if they are doing that, they wouldn't be here yet. But *tomorrow*," he adds with a smile, "they are going to arrive."

Perhaps the real answer is that interstellar travel is impossible, or at least impractical. Interstellar space might be filled with debris too small to be seen, but big enough to destroy a spaceship. Many astronomers think there is some *missing mass* in the universe, needed to explain the observed motions of galaxies. If some of that missing mass is in the form of small interstellar objects, space travel may simply be too hazardous. As Drake puts it, "You go flying through the galaxy at half the speed of light and meet an iron basketball and you've got serious problems."

The answer Drake favors is economic: "When you look at this in detail, it turns out to be be so *enormously* expensive that an in-

"I'm afraid this will be our last trip here for a while . . . We've had massive budget cuts."

telligent civilization wouldn't do it. A dumb one to be sure, not an intelligent one. And let me give you a feel for that, and you'll again recognize that here we are trying to psych out what other extraterrestrial civilizations and creatures might do—a very dangerous enterprise, since we can't even psych ourselves out. And here we are psyching out what intelligent spiders do on Beta Centauri or something. It's dirty work, but somebody has to do it."

Drake illustrates this with a simple hypothesis: an intelligent civilization wouldn't embark on an interstellar colonization unless the energy required to do it were no more than what's required to give themselves a good life on their home planet. So suppose that each creature is given a ration of energy—a "birthright"—for food, transportation, heating, and so on. Now, combine the rations of all creatures in that civilization to make up the energy of an interstellar spacecraft's motion. That's the basic assumption, Drake says, and it simplifies things, because we don't have to know how the propulsion systems work, and how efficient they are, or any of that sort of thing. In fact, the calculation yields the maximum speed of the spacecraft.

Looking at our civilization, if you take the average energy consumed per person in the United States over a hundred years, and assume ten tons of spaceship per colonist, you find the spacecraft cannot travel faster than sixty miles per second, 0.03 percent of the speed of light. First, Drake considered the case of a spacecraft the size of a Boeing 747 jet, traveling ten light-years. "Now get the picture," he said. "You are sitting with perhaps three or four hundred other people in a 747 airplane. For forty thousand years. Eating TV dinners. And so for most people this is unacceptably long."

Or consider a more realistic case, a hundred-year trip. The energy needed per colonist is enough to support 200,000 people in the United States. If you really need ten times that energy, to allow for inefficiencies, then the energy required for a hundred-person colony would be the same as that needed for the entire population of the United States for their lifetimes. "And there's no getting around this. There are no superefficient propulsion systems or such that will beat this game."

And furthermore, if you calculate the energy to generate colonies in nearby space—that is, in our planetary system—the energy spent turns out to be not something like 50 percent less but *ten million* times less. Ten million times less to establish the same colony in orbit within the planetary system. "So," concludes Drake, "this is, in my opinion, the answer to why they haven't come to Earth: Biology, physics and interstellar distance conspire to make interstellar colonization undesirable. What Congress would vote such a project that could run a country for a hundred years, for a mission whose outcome you will never know?"

Of course, if there are as yet unknown sources of energy, this argument may not apply. Or if we can harness solar energy and beam it as well as Robert Forward proposes, it may be that a civilization can divert just a fraction of its energy to send colonists out.

Carl Sagan has given perhaps the most outspoken reply to Tipler's anti-SETI arguments: "A lot of discussion seems to be dependent on the idea that beings *far* in our technological future will have motivations very similar to our own. We are asked to imagine enormous progress in their knowledge of and advances in the physical sciences, but we're also asked to imagine that in the biological sciences, and questions of the construction of their cultures, and sociological questions and ethical questions, they are as fundamentally backward as we are. So, for example, this *mad* notion of beings *strip-mining* the galaxy, running from world to world, tearing it up in order to make *more* strip-mining capability is, I imagine, from an extraterrestrial perspective, an indication of how backward we are on Earth, that we consider this a reasonable possibility, but not an activity advanced beings will find extremely unattractive."

SETI scientist N. L. Cohen of Boston wrote a tongue-in-cheek reply to Tipler in the magazine *Physics Today*:

> How do we know Frank Tipler exists?
>
> Have you ever seen Frank Tipler? . . . There are only 4 [billion] people on this planet; surely an intelligent creature would find some direct way of making his presence known to at least a sizeable fraction of the population.
>
> Perhaps we haven't seen Frank Tipler because we haven't looked hard enough . . .

What Now?

John Ball, of the Harvard-Smithsonian Center for Astrophysics, who proposed the Zoo Hypothesis, once summarized the "spectrum of possibilities" that confront us in SETI. He sees ten possibilities:

1. There are no other civilizations.
2. There are other civilizations, but they are primitive. They don't know about us, but might like to.
3. There are other civilizations, but they are at roughly our level. They suspect we may exist, and would probably like to talk with us. (This he calls the Mirror View.)

4. They exist, they know we're here, and would like to talk, but haven't managed to attract our attention yet.
5. They don't care about us. We pose no threat and don't have anything they want.
6. They're somewhat interested in us, and a few of their scientists are quietly studying us now.
7. They're very interested in us, and they're studying us extensively but secretly.
8. They are dabbling in our affairs right now.
9. We are an experiment in their laboratory.
10. God exists. (This is not necessarily inconsistent with any of the other options.)

It seems to me that the proper attitude is the one that has served science so well: accept that we don't have all the answers, and look for data to support one hypothesis or another.

Even if the pessimists are correct about life being a rare event, we should not discontinue SETI. Rather, we should just look at other galaxies. Our Milky Way is just one out of billions of galaxies in the universe. Even if intelligent life only evolves less than once per galaxy on the average, there still could be billions of other civilizations out there. A petition supporting SETI, initiated by Carl Sagan, was signed by about seventy of the world's leading scientists, including seven Nobel prizewinners. Some of them are skeptics, but all of them agree that SETI is worth trying.

The controversy brings to mind Arthur C. Clarke's First Law: "When a distinguished but elderly scientist states that something is possible, he is almost certainly right. When he states that something is impossible, he is very probably wrong." All of the scientists who signed the petition are distinguished, even if not all of them are elderly, so Clarke's law says we should pay more attention to the petitioners, and less to people like Tipler who say intelligent life cannot exist elsewhere.

The vigorous debate that the arguments of Hart, Tipler and others have stimulated is healthy. It would be unfortunate if SETI were to slip into dogma, because the history of science is one of dogmas being periodically overthrown. The universe doesn't care what we believe. The anti-SETI arguments have forced us to re-examine our positions and pointed out possible weaknesses. One good outcome of the debate, it seems to me, is the need to understand better the details of the origin of life on Earth and improve our understanding of the probability of each little step from mud to DNA.

But the ultimate point is that we just don't know what the probabilities in this story are. We only have intelligent "guesstimates."

Until we do SETI thoroughly, we won't know who is correct. We are in the position of the Arab camel-driver who listened to philosophers arguing about the number of teeth in a camel. Each philosopher believed in a different number, and argued endlessly about the philosophical reasons why the number had to be so-and-so. Disgustedly, the Arab turned to his camel, opened its mouth and counted its teeth.

The way to count the SETI camel's teeth is to build a receiver designed for SETI, and hook it up to a big antenna.

At worst, we won't find extraterrestrials, but will do good astronomy.

At best, the sky is the limit.

The Final Frontier

THE FUTURE

WHAT WILL HAPPEN if SETI is successful? Suppose, one day, we are watching television pictures of an alien civilization on the evening news, perhaps with a soundtrack translated with the aid of pictures, mathematics and physics. Each day might bring new stories from billion-year histories of alien cultures.

The effect on our society might be akin to the shock experienced by primitive human cultures when they came into contact with technologically more advanced ones. For example, in 1882, *Century Magazine* published the story of the Zuñi Indians' encounter with modern civilization.

The Smithsonian Institution had sent Frank Cushing to study the Zuñis of Arizona and New Mexico. He lived among them for three years and then brought some of them to Washington to meet President Chester A. Arthur.

The Indians traveled to Washington by the state-of-the-art in transportation technology: the railroad train. Several of them had never even seen a locomotive, and the experience of actually traveling like an arrow across the country made a powerful impression. As you read this, imagine that you have just stepped into an alien spacecraft, and you will better sympathize with their feelings:

> As they settled into their seats in the passenger coach they breathed a long sigh of gratitude, followed by their exclamation of thanksgiving, "E-lah kwa!" When the train started they raised the window-sash and prayed aloud, each scattering a pinch of their prayer-meal, com-

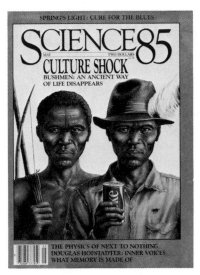

One example of the shock caused by a civilization coming into contact with a technologically more advanced one: the Bushmen of the Kalahari region of South Africa, many of whom have gone from Stone Age to high-tech in one generation. Will we experience a similar transition?

posed of corn-meal with an admixture of finely ground precious sea-shells, which they always carried with them in little bags.

During the afternoon they passed the pueblo of Laguna, at the sight of which they marveled greatly, saying, "Can it be that the sun has stood still in the heavens? For here in these few hours we have come to a place to reach which it used to take us three days upon our fleetest ponies!"

They inspect the alien technology:

At the water tower in Chicago they were awe-struck in the presence of the mighty [locomotive] engine, and became vexed with Mr. Cushing, because he prevented them from touching it, as they wished to, in every part, even where the action was most swift and powerful, with the thought thus to absorb its influence. "What if it should hurt us? It would nevertheless be all right, and just about as it should be!" said they, with their strange fatalism. They prayed before the engine, but not to it, as might have been supposed by some; their prayers were addressed to the god through whom the construction of such a mighty work was made possible.

They even encountered alien animals:

Driving though one of the Chicago parks they saw two sea lions, or walruses, which were kept there. Recognizing that they were ocean animals they almost broke their driver's arms in their impetuous haste to stop the carriages. They ran up to the animals, exclaiming: "At last, after long waiting, we greet ye, O our fathers!" considering them as "animal gods of the ocean," and began praying most fervently.

If an alien civilization is ever definitively detected by us, I fully expect some humans to start worshiping the aliens in much the same way.

The effect on our civilization would probably be overwhelming even with just the one-way receipt of information from them, without any physical visits. We can already see some of the possible effects on a small scale, when a nation receives television for the first time. Recently, for instance, the Central American nation of Belize began picking up American satellite TV broadcasts. Almost overnight, cultural habits of centuries changed. Life began to revolve around the TV set. People who had never heard of baseball became fans of the Chicago Cubs. The citizens will never be the same. They have absorbed an alien culture.

The signals from a more advanced civilization might contain the solutions to our greatest problems, problems that most likely occur to every civilization as it advances: dwindling natural resources, war, pollution, overpopulation, poverty, cancer. Solutions to these problems may be all around us, flying invisibly through the very room where the reader is sitting, just waiting for us to detect them.

If we don't find anything after searching very carefully, then perhaps the pessimists are right, and we are alone in the universe. But think of the tragedy it would be if we only listened to the pessimists and they were wrong! Signals might be waiting for us at this moment. They can be searched for so inexpensively.

But the search is getting more difficult, the longer we delay doing it. The world is becoming electronically noisier. Competition for space on the airwaves is so fierce that the sky is jammed with radio and TV stations, radar, military communications, telephone links and spacecraft. Ten years from now, it may not be possible to listen from Earth to most potential alien channels. Then, it may

One of the growing headaches of SETI: the crowded sky. Communications satellites such as these are pollution to SETI observers. Our own signals may prevent us from seeing E.T.'s. If we don't do SETI now, in a few years it may become impossible to do from Earth.

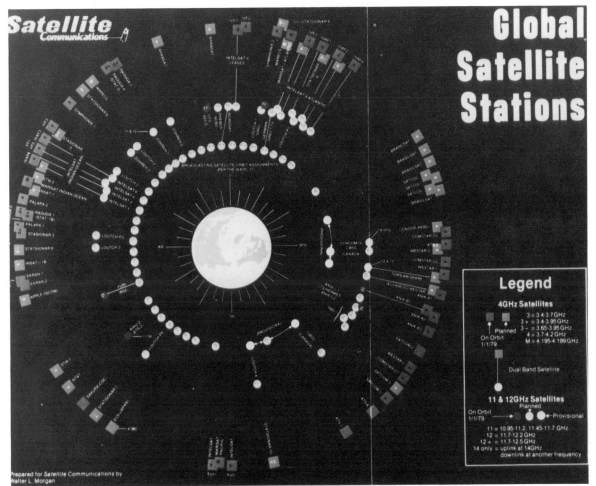

be necessary to go into space to do SETI. But that would be a vastly more expensive project, and consequently less likely to be done.

This is why SETI is not only important, but urgent. If we don't do it now, it may never get done.

What to Do?

What should we do in the future to increase our chances of success? I asked Frank Drake this question.

"We should do what we're doing," he said, "which is the NASA project on a grand scale, even more channels and a dedicated telescope, because you just have to cover many combinations of frequencies, places and times. There's no other way to do it as far as I can see."

Suppose there are means of communication that we have not yet even glimpsed? Perhaps we are as far off from the normal method of interstellar conversation as were the nineteenth-century thinkers with their giant triangles in the forest and their kerosene fires. This may not mean the search is hopeless, for some aliens may want to detect newly emerging primitive societies just as modern anthropologists search in jungles for lost tribes that have never been contaminated by contact with modern civilization.

The alien equivalent of anthropologists—or zoologists—might be the ones we first contact.

Alternatively, as I mentioned to Allen Hynek, I sometimes imagine that there are alien clubs similar to our ham radio operators, who

Project Cyclops, a long-range Stanford/NASA feasibility study, with hundreds of 330-foot dishes. If present SETI projects are unsuccessful, such a system could eventually be built, a few antennas at a time. It would be able to detect far weaker signals than now possible, would do superb conventional astronomy, and would be an excellent project for international cooperation.

delight at being the first to detect a new civilization, much as a short-wave listener jumps for joy when picking up a country he or she has never heard before. My alien clubs may have forgotten about technology as primitive as radio, but suppose they went to the museum or they saw it described in their encyclopedia, and discovered this thing that used to be called radio, the first technology their ancestors used to communicate over interplanetary distances, shortly after they emerged from the caves, millions of years ago.

My alien clubs consist of beings whose hobby is setting up and maintaining radio beacons, and monitoring for strange radio signals. If the universe is buzzing with life, this might be a particularly rewarding way to pass the time.

Carl Sagan, at the META symposium, considered the possibility of unknown technologies, looking beyond current SETI: "While we are just at the beginning of such searches, some people have from time to time asked if this is absolutely open-ended. What constitutes a failure in the radio search for extraterrestrial intelligence? Or will radio astronomers and physicists always invent a still more clever way in which the extraterrestrials might be signalling us—a mode in which we have not yet listened on—and so we never give up?

"Some people consider this a paradox. I do not myself. I think it makes sense not to give up. But, of course, the longer we listen on the greatest variety of systems without success, the rarer and more precious life here becomes, because it then would become increasingly apparent that life *that wishes to communicate with us*, at least, is not an absolute cosmic commonplace.

"From that point of view, it's natural to wonder about what other communication schemes there are. That is, we are *utterly* dependent, in Sentinel and META for example, on the intentions of the putative extraterrestrials. We are imagining that they wish to send us a message, and that they are, as Phil Morrison says, doing everything they can to make life easy for us. They may have means of communication that are far in our technological future, which we have not even glimpsed. I don't know what they are—zeta waves, say. And, if they wish to communicate with us using zeta waves, we, being extremely primitive and foolish, would not recognize that we are being hailed."

Practical ways around our potential ignorance are in short supply. But when I asked JPL's Thomas Kuiper what we should be doing, he had an unusual idea that could overcome at least some of the limitations of current SETI. He has a fresh approach to the search for the needle in the cosmic haystack.

"The haystack analogy," he pointed out, "is that you can go into the haystack and you can throw away everything that is not long and thin and silver, or you can take the approach that you throw away everything that is long and thin and yellow-colored, and look to see what's left."

He would like to see two radiotelescopes, separated by thousands of miles, that could stare at the sky looking for any kind of signal. Then, periodically, tape recordings from the two receivers would be compared to see if there were any unusual signal picked up by both. This would eliminate most sources of earthly noise, with the possible exception of communication satellites having beams broad enough to hit both telescopes. It's a wonderful idea because it doesn't assume anything about the type of signal other than that it's radio. It might also discover new astronomical phenomena that are currently ignored because they don't repeat regularly or don't look like the signals astronomers expect from the sky. Unfortunately, no one is planning such an observation presently. Perhaps this is an area for amateur SETI.

What Next?

There are several exciting projects currently under way, including the two that the Soviets are building. But the most important SETI development of the next decade will probably be NASA's. They are

JPL SETI scientist Edward Olsen in the RFI (Radio Frequency Interference) van used by the NASA SETI project. Here, scientists discover the types of noise SETI must contend with, and they test experimental designs of SETI receivers.

Part of the NASA SETI project in northern California, built by NASA's Ames Research Center and Stanford University. The power unit, computer, memory, and SETI spectrum analyzer are sometimes affectionately called the trash compactor, washer, dryer and refrigerator.

building the system originally opposed, but now agreed to, by Senator Proxmire. If it continues to be supported by Congress, it will be by far the most powerful SETI system ever. Like META, it will have around 8 million channels, but META can listen to only 8 million ultra-narrowband frequencies clustered around a single frequency. Thus, META can look at frequencies near the hydrogen line (1420 MHz) one year, and those near the hydroxyl (OH) frequencies (around 1700 MHz) another year. But the NASA system is designed to listen to broader bands so that it can eventually cover all frequencies between 1000 and 10,000 MHz. If the aliens are not using magic frequencies, META will miss them—but NASA may not.

Frank Drake described the capability of the NASA system this way: "We're getting to the point where we will address the thousand light-year frontier. Within a thousand light-years, there are ten million stars. So we'll have passed the point where the probability is on the order of fifty percent that we should have seen something."

Now, you think, what's the easiest thing to detect? Your intuition tells you that what you should search for is the nearest civilization. The nearest one should be the easiest one to detect, right? That's good common sense. But if you look at the history of astronomy, that particular intuition has always been wrong. The easiest stars to detect when you go out at night and look are not the closest ones but the inherently brightest ones which are very rare and in fact the farthest away.

A NASA maser used to detect the faint noises of the universe. A ruby amplifies microwaves the way a laser does light.

Drake points out that it's the same with radio sources. The easiest sources to detect are quasars, which are the most distant things in the universe, and also the brightest. Even though there are only a few bright ones, compared to many dim ones, those bright ones may actually outshine the faint ones.

"Now this could be the same with civilizations," says Drake. "And that creates a real problem, because your first intuition is that the SETI searches should start at the nearest star and then work outwards. That's your best chance. But if civilizations are distributed in their brightness like stars and radiogalaxies [galaxies with strong natural radio emissions], that's the wrong strategy. You should look directly at the most stars at once that you possibly can."

Since we don't know the answer to that, the NASA program is taking a two-pronged approach. They are looking at the whole sky, because perhaps there is one real bright one out there, and they concentrate on the Milky Way because it's particularly crowded with stars. But NASA also has "targeted search," where they look at the nearby stars, just in case most civilizations are about the same inherent radio strength.

If Congress grants operating funds, the NASA project should start full-scale observation in the late 1980s. Until then, META is the most advanced operational SETI system in the world.

In the longer run, I expect much work to pin down the terms in the Drake equation. Several space- and Earth-based instruments will give us far better views of the universe. The Hubble Space Telescope, a 94-inch diameter instrument will be launched by NASA, becoming our first large optical astronomical telescope in orbit. It will give us an unprecedentedly clear view of the universe. The Keck Observatory, which will be built on a mountaintop in Hawaii by Caltech and the University of California, will have multiple mirrors making it the equivalent of twice the diameter of the great Mount Palomar telescope, with four times the collecting area and much better visibility than from light-polluted Palomar. One day, we may go to the far side of the Moon to do astronomy and SETI.

The biggest uncertainty in the Drake equation seems to come from not knowing the precise steps that turned simple, lifeless molecules into complex ones that could duplicate themselves. Genetic engineering is making great progress in understanding the DNA molecule at the center of all life on Earth. Laboratory simulations of the primitive Earth's atmosphere and ocean are testing theories of molecular evolution and becoming increasingly more sophisticated.

But the wild card in this game, I suspect, is computer chemistry. Our understanding of chemistry and the physics of simple

Mars as seen by the *Viking* orbiter. Bright clouds of water ice fill the canyons of Noctis Labyrinthis, the Labyrinth of the Night. This is one indication that water could be available to microbes if any live in the Martian soil.

Seeing the invisible: a radar picture of the surface of Venus, made by the *Pioneer* Venus orbiter, showing a vast, mountainous continent. Perhaps an ocean existed in the lowlands during the early days of the solar system. (The "hole" in the bottom is not real; it is caused by lack of radar data.)

The surface of Mars, from the *Viking 2* Lander, showing a faint frost of water ice or dry ice. A small jettisoned cover lies near several trenches dug by the sample arm.

A *Voyager* close-up of Jupiter's Great Red Spot, a permanent storm large enough to hold three Earths side by side.

Voyager pictures of the four large moons of Jupiter. (Top, Io and Europa; bottom, Ganymede and Callisto.)

An artist's conception of the *Galileo* probe entering Jupiter's atmosphere in the early 1990s.

An artist's conception of Neptune, as seen from its mysterious moon, Triton, which may have lakes or oceans of liquid nitrogen. Its surface may be covered with an organic sludge.

A group of *Voyager* pictures of Saturn and some of its moons. In the foreground, from far right: Tethys, Mimas, Enceladus and Dione; in background, Rhea, and above it, Titan.

molecules is so advanced, and our computer programs so powerful, that we are beginning to be able to do chemical experiments without chemicals. We can sit comfortably at our computer screens and watch what happens when atom A meets atom B. And we can watch it in slow-motion, a billionth of a second at a time if we wish. Or we can do instant replay, or change the conditions and re-do the experiment.

Enormously powerful supercomputers are becoming widely available to the scientific world, making exceedingly complex calculations routine. And not only is the cost of microelectronics coming down each day, but new techniques of using dozens, hundreds or thousands of microcomputers simultaneously are now available in laboratories. Using such techniques, Caltech scientists have built the equivalent of a supercomputer that sits on a desk with room to spare, and costs a tiny fraction of the price.

Not only may these supercomputers help SETI to sift more quickly through the cosmic haystack, but they will also allow increasingly sophisticated organic chemistry to be modeled. I expect that one day we will be able to test many of the theories of life, without being hindered simply because we don't have the millions of years in which to experiment that nature has had. We will compress those millions of years into days on a supercomputer.

A possible Mars Rover robot. It could land at a safe spot and roam to interesting places where signs of past and present life could be searched for. The Soviets have recently expressed interest in a joint project to send a robot spacecraft to Mars that would return a sample of the planet back to Earth, possibly as a prelude to sending astronauts there.

We will simulate the primitive Earth, trying different possible paths of evolution, until one day, we will watch with our own eyes as the chemical Adam meets his molecular Eve . . .

Gravity Lenses

There is a remarkable idea just now surfacing in the SETI community that could eventually lead to a radically improved technique: gravity lenses.

The idea goes back half a century to Einstein. His general theory of relativity contained the first successful new theory of gravity since Isaac Newton watched his legendary apple fall. Einstein predicted that the gravitational field from any object would bend light that passed near it. The effect is too tiny to see under ordinary circumstances, but he predicted it would be seen in an eclipse of the Sun, when the Moon blots out the solar disk for a few minutes. When astronomers measured the positions of stars in the sky during an eclipse, sure enough, they found that stars' positions shifted slightly when their rays passed near the Sun.

The Sun acts as a lens, focusing starlight to a point far beyond Earth. It was predicted that galaxies sometimes might similarly act as a lens, focusing toward us by chance some more distant objects. In recent years, such gravitational lenses have been found.

These gravitational lenses are actually distant galaxies, and what they show are quasars, which seem to be even more-distant, violent galaxies. Astronomers have found several cases of quasars sitting far behind galaxies, their light bent by the galaxies so that we see two or more images of the quasar. These are intergalactic mirages.

The Sun, though it too acts like a lens, is too close for the effect to be of direct use. Light from another star would be focused 550 times farther away than the Earth's orbit, more than ten times Pluto's distance. However, the gravitational lens has the nifty property that it remains essentially in focus from that distance to infinity. This means that any star other than the Sun could be focusing onto us faint images of other stars.

Frank Drake described to a Caltech audience how this effect might be harnessed for SETI, an idea suggested by scientist Von R. Eshleman of Stanford University. "What does this look like? If you're looking down the line of sight, you're an observer out along the focal line, you see a ring of light. If you're slightly off the line connecting the two stars, the ring breaks up into two segments, and you get just a little more away from that simple line and see just a normal stellar image."

The nearby star acts as a lens, but an incomplete one—because the star itself blocks the direct rays of the farther star which you are trying to see. Only the light bent around the outside of the star reaches us.

However, the light-gathering ability of the lens is extraordinary. On Earth, optical astronomers work hard to get as large a collecting area for their telescopes as possible. Thus the Mount Palomar telescope with its 200-inch diameter mirror has an area of around 200 square feet. The effective light-gathering area of a star similar to the Sun would be a ring around the star with an area of billions of square miles.

"So," says Drake, "one has collecting areas here which boggle the mind. Besides that, the sharpness of the image turns out to be enormously great. The diameter for instance, at the minimum distance of the Sun is only one hundred meters [about 100 yards], reflecting the enormous sharpness provided by this large lens. If we could but see it, this is what every star in the galaxy looks like, sort of a sea urchin if you will, a star making images of every other star, starting at the minimum distance and going out to infinity. This casting of very high-resolution images of the whole universe on the sky—and these images are in focus at all distances—is a really remarkable thing."

Each star produces hundreds of billions of tubes of light, one for every other star, so there are hundreds of billions of hundreds of billions of tubes of light in our galaxy.

The gravitational lens also focuses radio signals in the same way. Suppose we use a ten-meter antenna on Earth. (For radio astronomy, that's small; for optical astronomers, that's the world's largest telescope.) We would have a collecting area equal to that of a dish eleven kilometers in diameter. That's equivalent to a *thousand* Arecibo telescopes, using just a tiny antenna. And if we could use Arecibo itself as a collector, we'd get a collecting area equivalent to a million Arecibos, for both transmitting and receiving.

What that implies is that we can detect minuscule signals all the way across the galaxy just by using a gravitational lens. We can also create extremely powerful signals at distant points by using such a lens. In fact, we could put an ordinary TV transmitter on a telescope like Arecibo and use another star's gravitational lens, broadcasting television in its full glory all the way across the galaxy—if that's what we wanted to do.

The image produced this way can show details as small as ten miles wide. "A startling number. That's much less than the size of a planet, which means that if you use gravitational lenses as a telescope, you can not only detect planets, but actually observe

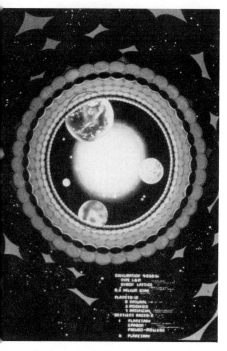

An imagined excerpt from the Encyclopedia Galactica we may one day decipher if SETI succeeds.

details on them, and perhaps find life even more primitive than our own.

There are serious practical problems with this approach, but it is a radically new technique to be studied for the near future. There seems to be almost no limit to what human ingenuity can devise, which makes it all the more likely that some technique will succeed in detecting extraterrestrials in our lifetimes, if they exist.

"I suspect," concludes Drake, "that civilizations that are more competent in space-faring than we are use this device—use it to study the cosmos, in the ultimate telescope, perhaps also to send transmissions to other civilizations. It's quite possible that through the use of gravitational lenses, every word, every transmission we make is detectable by the remotest stars."

First Contact

Suppose that someone, somewhere, finds that Big Clue, the one that leads to the first indisputable proof that we are not alone. Excitedly, he (or she) shares it with colleagues. At first he is greeted by skepticism, but gradually eyebrows start to raise as, one by one, they cross off the list of possible natural phenomena. Then confirmation comes in from other observatories. We have First Contact!

What then?

Some people argue that if the civilization is thousands of light-years away, we can't have meaningful conversations with them since it would take thousands of years for our "hello" to get back to them. But history provides examples of one-way communications over thousands of years, and some of them still affect our lives even today.

Most of the world's major religions represent such communications. Books written centuries ago such as the Old Testament, the New Testament, the Koran, the Bhagavad-Gita, the writings of Buddha and Confucius, shape the lives of people now, even to the point of causing wars.

The ancient Greeks and Romans have shaped the course of modern civilization in many ways, both good and bad. Their influence shows the effects of a one-way transmission of information over centuries of time.

Many ancient Greek writings, however, were lost to Western civilization for centuries. They survived only because the Arabs preserved them. The re-introduction of those writings from Arabia into Europe provided a culture shock that may hint at the potential impact of the decipherment of alien signals.

In the twelfth century, there was a medieval Italian scholar called

Gerard of Cremona. He was a man dedicated to knowledge, who spent most of his life at the college of translators in Toledo, Spain. Year after year, he pored over Greek and Arabic scrolls, a researcher with a quill pen, buried in the equivalent of today's computer print-outs. He translated these documents into the standard language of the educated European, Latin.

The science of the Arabs was then the greatest in the world, and its value was magnified by the translations that Arab scholars had carefully made long ago of the writings of Greek scholars, often from the originals, sometimes from Syriac and Hebrew translations. Much of the original Greek literature eventually perished, and but for those Arabs, it would have been lost forever.

Through Spain—then under Arab occupation—this ancient culture flowed into Europe. Knowledge of the most advanced medicine, mathematics, astronomy, astrology, alchemy and philosophy became available there for the first time. This was a turning point in European history. The light of knowledge began to illuminate the superstitions and ignorance of the Dark Ages.

Gerard had been attracted to Toledo by his love of the greatest work of ancient astronomy, Ptolemy's *Almagest*, a Greek work that survived only in Arabic. In Spain, he found a vast storehouse of knowledge lost to Europe, and he learned Arabic in order to translate it. The most prolific translator of his era, he translated not only the

A hypothetical alien civilization built around its planet.

Almagest, but other writings of Aristotle, Euclid's classic of geometry, *Elements,* a tract of the great engineer Archimedes, the works of the great physicians Galen and Hippocrates, and many other works on astronomy, algebra and optics. In *The Renaissance of the 12th Century,* historian Charles Homer Haskins says, "Indeed, more of Arabic science in general passed into Western Europe at the hands of Gerard of Cremona than in any other way."

These writings revolutionized European culture and made possible the later intellectual revolutions that have formed the modern world.

Passing from history into the present, here are some modern visions of the possible effects of our detecting alien civilizations, starting with Thomas Kuiper's startling reply to my question: What do you think would be the consequences of first contact?

"Very little," he stated bluntly. "And the reason for that is that we won't know that we've had a contact." He thinks that interstellar beacons or other signals intended for us are unlikely. "Our best strategy, I believe, is to search for emissions arising as byproducts from alien activities."

How will the first contact come about?

"It depends on how you mean a 'contact.' If you mean the first *detection,* it's even conceivable that we might already have done it, but simply don't recognize it. I envision that turning detections into an awareness of the existence of another civilization will be somewhat analogous to the process whereby we've come to accept the existence of black holes. I don't think we can yet say with utter confidence that black holes exist, but I don't think that there are very many astronomers left who don't believe that they exist, and the confidence level is pretty high that certain objects in the sky are in fact black holes. And it took a long time to get to that point. Basically, we saw an accumulation of bits and pieces of evidence."

So Kuiper feels that if we do come to recognize the existence of an extraterrestrial civilization, it will be the result of an accumulation of phenomena that are hard to account for. For any single phenomenon, he says, we can almost always find some kind of a natural explanation. It's only when we start accumulating a *variety* of phenomena and the collection of natural explanations becomes too strained that we will begin to believe that perhaps a consistent explanation in terms of extraterrestrial civilizations is the most plausible. And such a process would take a long time, so its social impact would be diffused over time. According to Kuiper, we will have generations to get used to the idea.

"Of course," he acknowledges, "it will still have some impact over that period of time. I think it'll change our attitudes considerably towards ourselves, our fellow men, the universe at large. But it'll be a *slow* process, not suddenly reading in the *New York Times* that extraterrestrial civilization has been discovered, and what are we going to do about it? I don't see that scenario as very likely."

Perhaps he's right, but many of us expect that such a discovery will be much more sudden, more like the discovery of pulsars: some phenomenon in the sky will be detected that seems unnatural. Scientists will excitedly focus on the signal and prove it is extraterrestrial. Then they will torture their brains to think of any natural process in physics that could conceivably cause such a signal. They will also have to test for hoaxes. (At one SETI meeting, several scientists discussed how to avoid being the victim of a Caltech student prank.)

Then, after they have convinced themselves that contact has at last been made, would come the announcement to the world.

Headlines would banner the announcement as they did the first human landing on the Moon. The world's media would converge on the observatory, and the scientists responsible would be credited with the greatest discovery in history.

Here is SETI pioneer Philip Morrison's view: "I think that the feeling people get by thinking about it a while is something like this: it'll be a sensation when it's confirmed. In fact, it will be quite a sensation *before* it's confirmed. And then it'll turn out to be false. And then finally, it'll be confirmed and that'll be a sensation. It'll occupy the front pages of all the newspapers for a week, and occupy the inside pages for another month. And then it'll occupy a weekly piece every once in a while for six months, and that'll be the end of it. But then, scholarship and interested people will emerge everywhere.

"As you know, the television cameras will come the very next day, and reporters will ask 'What do they look like?' And, 'Do they speak English?' and a number of other such things, projecting the provinciality of our own nationally divided Earth, and our own limited cultures, upon something so remote we have no way of judging it."

Interstellar-travel advocate Robert Forward thinks it would greatly broaden the perspective of the human race. "It would be a longer view," he says, "a longer vision of the universe, and give me more hope. I think—I hope—that's the way most of the world feels."

But might the discovery be hushed up? People often ask me that. They repeat rumors that the Air Force has hidden wreckage of flying saucers. They think of all those science-fiction movies where the first response of the government is to make contact with the aliens a secret.

META creator Paul Horowitz has a good response to that: "We're looking for beacons, not messages, and the best beacon is the worst message in general." A beacon would be so much stronger than an information signal because its main purpose would be to stand out from cosmic noise. It would be designed to attract attention, and once that had happened, a listener would automatically look for other, subtler signals from the same point in the sky, and *those* could contain sounds, mathematics, language lessons and television pictures.

"The beacon's not going to say a hell of a lot," said Horowitz. "It's going to say, 'We're here. We're transmitting at this frequency. We invite you to find our message channel.' Which is probably related— to maybe some modulation in the beacon that points to the message channel or whatever. Therefore finding the beacon—finding the signal—if and when it's found, will not reveal any kind of information in the national security interest or anything else."

But might the government try to keep any such signal secret because of the military implications if information about an advanced technology were then deciphered?

Horowitz thinks that it would be impossible to keep such a discovery classified, because in the process of verifying it, it is necessary to have scientists at other observatories look at the same place in the sky, just to make certain that you're not seeing an artifact of your own observatory. "By that time, the cat's out of the bag, because you've got a dozen people and their wives and kids and dogs and cats, who are in on the world's greatest discovery, and they're not going to sit on it. And it gets out. So, I think that you don't have to worry. It's going to spill out, it's going to be all over the place."

What might we find once we start deciphering the signals?

Horowitz speculates that if we find the beacon and it leads us to an information channel, and we start listening to that channel, we might find some sort of monologue that contains a language lesson—perhaps based on math and science—possibly interspersed with more complicated material.

We may receive messages having to do with, say, art and political science—things associated with the alien culture. Such information might not make any sense for decades or centuries, simply because they are so alien. "Perhaps in the first weeks or months— just spinning out the scenario for whatever it's worth—we might understand the symbols for plus and minus and equals, and start to get some handle on this message, but a blueprint for a rocket is not in the first century, I don't think, of decoding this thing. There will be departments set up at universities to decipher major messages. People will get their tenure that way.

"And," Horowitz adds, "there's been no interest whatsoever in this project, or any project like this, by the military, that I've ever heard of." Carl Sagan points out that the same is true for the invention of rockets. After a while they got interested.

Would there be much culture shock?

Robert Forward expressed a surprising view: "No. Somewhat, but again, it'll be a one week shock. People go on to the next amazing thing, a few weeks later."

Morrison agreed. "It will be a shock to some people, perhaps profound to a few. But on the whole, it's been so discounted by the elaborate imaginative structure of our times—science fiction and the rest—that I don't think that it'll be that much. In fact, it is my opinion that it'll be somewhat the other way around. I think that out there in the streets, quite a few people believe that the enterprise of radioastronomy, *as a whole*, is devoted to this search, and of those, quite a few believe it's already succeeded."

But wouldn't the discovery have profound effects on anyone? Sagan ruminated on this point. "Let's just for a moment imagine that news is out that a complex message has arrived. And then let's imagine the reaction of people in various occupations. Let's start with astronomers. A lot of astronomers will of course feel delighted because there'll be lots of astronomical data in there, but there'll be a little undercurrent. Have we missed something elementary? Will we discover that the textbooks are wrong, and maybe we're responsible for propagating misinformation? Just a *little* uneasiness, I think, that would happen about that.

"And now try to extend that to fields like social organization, politics, religion. I think that there will be a lot more disquiet in those fields. What happens if the extraterrestrial wisdom is totally different from what we consider the conventional wisdom here? I think that there is a significant potential for culture shock along *those* lines, in the same sense that even the discovery of Native Americans represented a much more mild, but still significant, culture shock, and notice it was used by European social critics like Voltaire, who imagined a Huron Native American transported to the court of France, who could then ask out of naivete, questions that were searching and fundamental, which nobody on the French court would be willing to ask themselves. I imagine a kind of serious reassessment of the conventional wisdom would follow just the knowledge that a complex message had been received, never mind the actual deciphering of the message.

"Basically, I think that the positive aspects, the benign aspects, the forward-looking aspects, the sense of entering a new era, *far* outweigh the potential negatives. And it illuminates the approach to central questions of our being: the search for *who* we are. Are we the most intelligent beings in the universe, or are we just a cosmic commonplace? It's hard to think of a deeper question than that, and it's very hard to answer that question—before we receive a message."

Having listened to so many scientists relate their ideas about first contact, I couldn't resist asking a poet and science-fiction writer who has created other-worldly dreams that millions have shared. Who better than Ray Bradbury to provide a different point of view? He's the next-best thing to an alien. Until one comes along, he'll have to do.

Suppose SETI succeeds?

"Well," he said, "it's just one more proof of how fecund the universe is. You know, if I were going to become a rabbi or a priest or a minister tomorrow, I would have every Sunday morning or every Friday evening nature films and films on astronomy because the universe is so damned miraculous no matter how you look at it, scientifically or religiously or whatever, that when you see the symbiotic relationship of some of your life forms, it becomes increasingly stunning how the universe has developed.

"So, therefore, if we did discover an intelligent life in other parts of the universe, there'd just be one more proof of how incredible our existence is. Of course, on every planet that's half-way decent

Seashell/galaxy: a metaphor of the universe.

$$r = a e^{\theta \cot \alpha}$$

in the universe, you're going to have tens of thousands of life forms unlike anything on Earth, and in fact, there will be no duplications. Nothing on Earth will be duplicated on any other planet, and the building blocks will be the same, but all the things that fly in the air will be different, all the things in the sea will be different and all things on the land will be different."

What would be the effect on human society today?

"It will make us more religious, as well as more scientific."

Religious?

"Well, the word *religion* means 'to relate yourself to something,' that's all the word means. It's based on the old latin term *religio*, to bind together, the binding of sticks, of taking separate facts and binding them together with a theory or a law, and the bundle of sticks that represented ancient Rome was exactly that."

And finally, I couldn't resist asking him: What would you tell E.T. if he called?

"That we're one and the same. That we share the universe. But we haven't learned to do that on Earth yet, have we, which is very sad. But we would love it, and we preach it to one another. Unfortunately, the people that run the world, which is not us, don't do any listening."

The Future

What if SETI is unsuccessful? What if, decade after decade, we observe in every way possible, and gradually come to the conclusion that we are alone?

That could happen, even if it seems very unlikely. It might mean that life is rare, and intelligent life even rarer. In the extreme case, we could turn out to be the only intelligent creatures in our galaxy.

We would have a galaxy all to ourselves. If we succeed in avoiding self-destruction, it will then be humans who will spread out into space.

Humanity has never left a niche open for long when it was discovered anywhere on Earth. However comfortable some people are in their homes, there are always others who are attracted to frontiers: adventurers, misfits, idealists, religious zealots, merchants, outlaws, the persecuted and those who seek unrestrained freedom. First, they will move to space stations around Earth, then to the Moon and Mars, then to asteroids and comets, and ultimately to the stars.

At that point, humanity will become indestructible. No war, no accident, no cosmic catastrophe will be able to destroy the entire human race. During the next billion years, the human species would, I am convinced, become the extraterrestrials for which we have searched so hard.

But what if SETI succeeds? The above expansion will probably occur anyway, though constrained by the knowledge and guidance of other civilizations. Many of the mistakes we might otherwise make would be avoidable if we just had the benefits of the history of older cultures.

The important lesson of SETI history thus far is that, above all else, SETI requires patience. It is not surprising that we have not yet succeeded. Most SETI projects have been very brief, and all have used limited equipment, money and personnel. Only a tiny part of the electromagnetic spectrum has been searched. We have just barely begun to look for the needle in the cosmic haystack.

Fortunately, there are many ways in which first contact could happen, even if the skeptics are right about interstellar travel being too difficult. One way, of course, is through the various SETI programs around the world.

A second possibility is from some conventional astronomical program, one scanning the sky for quasars or pulsars or infrared galaxies or some other member of the crowded astronomical zoo. One of the exciting things about astronomy is that it so often leads to unexpected discoveries. Perhaps an astronomer will find a star whose spectrum contains an element that does not occur naturally, or a gamma-ray source that cannot be from a black hole or other familiar object. Our first clue could be the byproduct of some industrial process we cannot even imagine: galactic pollution.

A third possibility is amateur SETI. Ham radio operators, electronics buffs, computer hackers and amateur astronomers have started a number of small-scale SETI projects. Robert Stephens' project in Canada, using surplus military radar antennas, may even rival some of the professional SETI projects of the past. Near Chicago, Robert Gray and collaborators have a "Small SETI Observatory" using a satellite dish, which is being used to search for a repeat of Ohio State's "Wow" signal. In Silicon Valley, California, ham radio operators are using a similar antenna and a microcomputer for SETI, with advice from Kent Cullers, a blind scientist on the NASA SETI project.

As satellite dishes proliferate and surplus electronics becomes available, the power of amateur SETI will grow. NASA and Stanford University are investigating the possibility of putting much of their SETI electronics onto integrated-circuit chips, and perhaps one day a company will sell SETI chips as inexpensively as microprocessor chips are now—a few dollars apiece. Perhaps some amateur, looking in a part of the spectrum ignored by the professionals, will make the discovery. (Former NASA SETI scientist John Wolfe points out that the band at 1000 MHz frequency and below tends to be relatively quiet, yet is often neglected by professionals; it's readily accessible to amateurs.) Perhaps someone using their ears or eyes to study noises from the sky will find a pattern so sophisticated or bizarre that the professional SETI computers overlooked it.

A fourth possibility is from communication engineering. Every day and night, engineers and technicians aim their microwave dishes at the sky, picking up signals from the communication satellites on which our radio, television, telephone and computers are increasingly dependent. One of their largest headaches is tracking down noise. The frequencies they use are often right where SETI scientists search for signals. Perhaps one such signal will turn out to be an interstellar beacon.

We should never forget that the entire branch of radio astronomy was started not by a professional astronomer, but by a radio engineer, Karl Jansky. In 1932, he published his discovery of signals from the center of our galaxy. His results were ignored by the

A picture from the 1930s showing the inventor of the radiotelescope, Karl Jansky, of Bell Telephone Laboratories. His investigation of strange telephone noise led him to discover natural radio signals from the center of our Milky Way galaxy.

astronomical community for a decade, even though they were hardly concealed: an announcement appeared on the front page of the *New York Times*. It was another nonastronomer, radio engineer Grote Reber, who followed up Jansky's discovery and built it into the modern science of radio astronomy. Perhaps to some such "amateur" will come the greatest discovery in history.

Yet another possibility is from nonastronomical science. Perhaps some great particle accelerator, miles in diameter, being used to study the structure of the smallest pieces of matter, may reveal an unexpected "glitch" that is a clue to a new way to communicate. Or some humdrum experiment in a laboratory may lead to a new way to communicate, which would mean a new way to listen. Perhaps a new Alexander Graham Bell will pick up the the first "telephone" and find that someone is calling . . .

A sixth possibility: perhaps contact will come about through the numerous military satellites that occasionally do inadvertent astronomy. X-ray-emitting objects in the universe and comets hitting the Sun were found this way, the information kept secret until it could be released without compromising security. Perhaps the first signs of an extraterrestrial civilization are already sitting on magnetic tape, deep in the bowels of the U.S. National Security Agency or the Soviet KGB. Perhaps they have been rejected as natural noise. But one day they might be studied because someone thinks they may be a new trick in the opponent's technology. Perhaps a James Bond will be the first to discover signs of another civilization.

Or perhaps the big clue is already here, but frustratingly unrecognized. It might be seen by someone patiently sifting through miles of computer tape from an old experiment such as IRAS, or by a student seeing some peculiarity in a graph in an astronomical journal whose significance no one had previously recognized.

And there is at least one more possibility. One day some astronaut might set foot on the Moon or Mars or an asteroid and find *something* unnatural—an artifact, a piece of debris from another civilization, one that may have drifted through the dark, airless lanes of the galaxy for billions of years.

Out of the many possibilities, which is the most likely? Let me share with you the scenario that I suspect is closest to what may happen.

Sometime in the next few years, one of the SETI projects around the world will, I expect, discover a strange signal. It will be something simple that cannot occur naturally: perhaps a pure, constant tone. Or the digits of the number pi, 3.14159 . . . , expressed in the most fundamental of all number systems, the binary system of zeroes and ones used by computers. Or it might be a series of prime numbers, numbers such as those used in Drake's simulated SETI picture, which cannot be divided by any number but one and themselves: 1, 2, 3, 5, 7, 11, . . .

As far as we know, such a signal could not be produced by anything but an intelligent civilization. This would be proof positive that the source was intelligent.

Excitedly, scientists will study it carefully from different observatories, to rule out the possibility of Earth-based interference, artificial satellites and hoaxes.

The Beacon will have been found.

Then the announcement will be made, and the world will go crazy for a while. Other scientists will greet it with skepticism, but when observatories all over the world confirm it, they will have to agree that we are not alone anymore.

The most powerful telescopes of every nation will focus on that point in the sky where the signal comes from. Every weapon in the astronomical arsenal will be turned on it: telescopes using radio, infrared, optical and ultraviolet wavelengths; X-ray, gamma-ray, cosmic-ray and neutrino telescopes; every astronomical satellite. More powerful instruments will be built as soon as possible, and money will be no object for the first time in astronomical history.

One of those instruments will discover a complex signal coming from the same point in space, and soon many such signals will be found. After much experimentation, it will be found that some of them are television signals, in color and three dimensions.

Every TV newscast on Earth will then show pictures of an alien being, looking like nothing ever seen on this planet even in the movies, a creature stranger than we can imagine, speaking an incomprehensible language.

All over the world, billions of people will stare in curiosity and excitement and horror and amusement and fascination at this strange creature on the TV screen. Suddenly, many people will begin to realize that the differences between Arabs and Israelis, whites and blacks, Russians and Americans, are utterly trivial compared with the differences between Us and Them. Ancient hostilities will not disappear overnight, but it will become harder to stir up old hatreds, and some people will be shocked into thinking about their place in the universe for the first time.

We will find many channels of alien signals, and one of them—distinctly different from the others—will be a dictionary. The signals will repeat endlessly, so it will not matter if we pick it up somewhere around the alien equivalent of the letter *M*. The dictionary will use pictures and the laws of nature to teach us their language, and perhaps how to build better receivers, or receivers of a type we have not yet dreamt of.

Most SETI scientists expect the translation to be extremely difficult, and to take years or even centuries to crack. I disagree. Consider this: the Earth is five billion years old, while the age of the universe is around fifteen billion years, so the odds are that the other civilization will be far older than ours. We have only been human for a few million years, and have had radio technology for about a century, so the chances of a detectable civilization being just at our level of technology but not much beyond are very slight.

If they are far ahead of us technologically, then they will be just as advanced in their ability to *teach*. They may have had thousands of experiences in teaching their language and culture to other primitive civilizations, and they now know how to do it very well—if they want to. (If they don't care about savages like us, then we may have no help in deciphering their signals, and *that* could be horrendously difficult—like Aristotle trying to decipher a copy of *Time* magazine.)

If they choose to transmit lessons, it may be easier to learn their basic language than it is for an English speaker to learn Chinese, although it could take a lifetime to adjust to the alien concepts. Probably only children who are raised in the post-contact world will truly understand the subtler aspects of alien culture. The rest of us will be too rigid in our thinking.

Some of the other channels will probably be an alien encyclopedia, repeating periodically, filled with the history, science, ethics, art, politics, music, morals, literature, religion, games, sports and entertainment of their worlds, plus those of other civilizations they have contacted before. There will probably be whole facets of their culture that have no correspondence to any subject we know—the poetry of equations, the esthetics of interstellar clouds, the etiquette of sharing minds.

There will be articles on curing diseases in carbon-based lifeforms, new sources of energy, how to transmute waste products into any desired element, popular spaceship designs. The two most important articles, I suspect, will be "101 Time-Tested Ways to Prevent Thermonuclear Holocaust," and "How to Prevent Interstellar Culture-Shock from Ruining Your Civilization."

Suppose they are religious? Their religion cannot possibly agree with all of the thousands of religions of Earth. Some humans will reject it on the grounds that they must be tools of the devil. Others will treat it with the respectful disagreement with which many in the Judeo-Christian culture treat Moslems, Buddhists, Hindus and others. Some people, I'm sure, would quickly adopt any alien religion.

Suppose they are irreligious? Then, again, some will reject them as tools of the devil. Others will, I am sure, worship the aliens as gods. If they are millions or billions of years beyond us, they will probably have godlike powers, just as we have powers today that are godlike compared with, say, medieval Europe: the power to communicate over vast distances instantaneously, the power to cure diseases, the power to travel halfway around the world in minutes, and the power to destroy whole cities in an instant. The aliens will probably have powers as far beyond those as ours are beyond Medieval Europe.

What about their politics? In the unlikely event that they have a familiar political or economic system, this would greatly alter current political conflicts. If their system resembles Fascism or Marxism or Western democracy, this will create powerful emotional support for the lucky system. Far more likely, I think, is that they will be as far beyond those current political labels as those are beyond the organization of a nomadic tribe.

After all, their culture will be based on a history very different from ours, and their biology will probably be at least somewhat different, too. What would our culture have been like if we'd been descended from cold-blooded reptiles like the dinosaurs? Dinosauroid beings might consider our history to have been extraordinarily cooperative and pacifistic. Or if they are derived from passive grazing animals like cows, they might not even have a word for "war." Or consider sex. Biologists still can't agree on why two sexes arose on Earth instead of just one. Maybe sex is an unusual phenomenon in the universe. A society without the struggle for sex-partners, and in which children are identical to their parents, would be totally alien to us—as we would be to them.

Even if they biologically resemble us, their social system will probably not even be relevant to us at the present stage of our history, any more than nomads need worry about Internal Revenue Service Form 4797. But people will begin to absorb bits and pieces of the alien system and try to adapt them to the rapidly changing human culture, in much the way that third-world nations have adapted some facets of European colonial cultures to their native ones. This time, however, the changes will be adopted voluntarily instead of being imposed by a foreign power.

It is conceivable that the aliens may send only a sample of what they have to offer, because their real goal may be to receive *our* encyclopedias, the only thing we have to offer a more advanced society: the unique products of human culture, the art and history that could never be duplicated on another world, no matter how vast the universe. There could even be an interstellar economy in which the sole currency is information, being traded back and forth among diverse civilizations at the speed of light (or faster, if they have a way around this speed limit).

More likely, I think, they will just send a complete encyclopedia, along with a description of a clever way to build a powerful transmitter, plus a request that we transmit to them our own encyclopedia. There would be suggested frequencies and methods of operation to prevent interference with signals of other cultures: Miss Manners' *Book of Interstellar Etiquette*.

The United Nations will convene to decide how to handle the response. There will be a great debate about whether we *should* reply. Some suspicious people would resist on the grounds that we should not give away any information to a potential enemy. Some would want only a censored encyclopedia transmitted, one that omitted the countless embarrassing and horrifying parts of our history, to make us appear more civilized than we really are.

But we have already given away the deepest secrets of our culture in our TV transmissions. It is too late to hide our ugly side from the galaxy. Furthermore, the aliens would be so far beyond us that we would have nothing to lose and everything to gain. The temptation to thereby transmit our heritage and receive a kind of immortality would be irresistible.

I also envision a project in which every human being is videotaped and their names and faces transmitted into the galactic storehouse of knowledge, giving every person on Earth a piece of interstellar immortality more enduring than the pyramids of Egypt.

Something close to literal immortality may result, too. The medical knowledge that an advanced civilization must have, if applicable to our own biochemistry, would enable us to vastly extend our life expectancy. The likelihood is that they have encountered so many lifeforms that, even if their biochemistry is different from ours, they know the general principles of all biological systems. They would probably tell us how to decipher genetic codes atom by atom, and to routinely fix the errors we call diseases. Death by natural causes would become unheard of. Only death by accident, crime and war would still be possible, and those might be largely eliminated by alien wisdom as well.

This could make even worse the already horrendous population problem of our planet, but the medical knowledge would probably make birth control trivial, and access to the resources of the whole solar system would allow us to survive with a much higher standard of living for everyone, even with a far larger population.

This planet would never be the same again, and, eventually, we would be ready to join the universal culture of *civilized* creatures.

This is what I suspect will happen. I also expect to be surprised by discoveries that no human can possibly predict, for such has always been the history of the exploration of the universe.

With SETI, we are treasure hunters, scouring the sky for a treasure greater than the golden cities sought by the Conquistadores. Where treasure hunters today use metal detectors in hopes of finding a lost coin or a pretty artifact, we are searching the sky for a treasure-trove of knowledge—at the very least, the knowledge that we are not alone. We have no native guide but the laws of physics, and no treasure map but the astronomers' surveys of the largely-unexplored universe, filled with regions marked "quasars" and "pulsars" and "black holes," evoking the "here there be dragons" of the ancient sailors' charts.

One day, we may watch eerie scenes of alien television programs, as in this computer graphic. (Ned Greene, NYIT)

We have whole celestial continents to explore. At the end of that exploration, we will certainly know far more about ourselves, our origins, our destiny. And just possibly, we will find the solutions to our Earthly problems.

Never before in history have so many deliberate searches for signs of extraterrestrial civilizations been conducted. Never before has such powerful technology been harnessed. Never have such advanced non-SETI astronomical programs been undertaken. There is no way to be certain which strategy is best, so the mere diversity of approaches being used by so many people for so many purposes increases the chance that one of them will be the right one.

Someday soon, I suspect, we will find that we have strange-looking friends on worlds more exotic than we can imagine.

I think intelligent life probably exists elsewhere. If it does, there is a very good chance we will detect the first signs of an extraterrestrial civilization before this century is out.

If E.T. is phoning, *we* are listening.

SUGGESTED READINGS

Chapter 1. E.T.: Phone Earth!
The Nature of the Search

Angelo, Joseph A., Jr., *The Extraterrestrial Encyclopedia*, 1985, Facts on File, New York. Fairly technical book on all aspects of life in space.

Asimov, Isaac, *Extraterrestrial Civilizations*, 1979, Crown Publishers, New York. Nontechnical.

Audouze, Jean, and Israel, Guy, *The Cambridge Atlas of Astronomy*, 1985, Cambridge University Press, New York.

Baugher, Joseph F., *On Civilized Stars*, 1985, Prentice-Hall, Englewood Cliffs, NJ. A recent, nontechnical look at SETI.

Bracewell, Ronald N., *The Galactic Club*, 1975, W. H. Freeman, San Francisco. Nontechnical.

Cantril, Hadley, *The Invasion from Mars*, 1966, Harper and Row, New York. The true story of the panic from the radio broadcast.

Christian, James. L. (ed.), *Extraterrestrial Intelligence*, 1976, Prometheus Books, Buffalo. Thoughts of scientists and science-fiction writers.

Edelson, Edward, *Who Goes There?* 1979, McGraw-Hill, New York. SETI. Nontechnical.

Feinberg, Gerald, and Shapiro, Robert, *Life Beyond Earth*, 1980, William Morrow, New York. Nontechnical.

Goldsmith, Donald, *The Quest for Extraterrestrial Life*, 1980, University Science Books, Mill Valley, CA. A fascinating collection of most of the classic scientific papers on SETI. Most are technical, but many are quite readable.

Goldsmith, Donald, and Owen, Toby, *The Search for Life in the Universe*, 1980, Benjamin/Cummings, Menlo Park, CA. Nontechnical.

Morrison, P., Billingham, J. and Wolfe, J., *The Search for Extraterrestrial Intelligence*, 1977, NASA SP-419; reprinted by Dover Publications, Mineola, NY. Technical.

Papagiannis, Michael D. (ed.), *The Search for Extraterrestrial Life: Recent Developments*, 1985, D. Reidel, Boston. Technical.

Ponnamperuma, Cyril, and Cameron, A.G.W. (eds.), *Interstellar Communication: Scientific Perspectives*, 1974, Houghton Mifflin, Boston. Semi-technical.

Ridpath, Ian, *Messages from the Stars*, 1978, Harper & Row, New York. Nontechnical.

Rood, Robert T., and Trefil, James. S., *Are We Alone?* 1983, Charles Scribner's Sons, New York. Nontechnical.

Sagan, Carl, *Cosmos*, 1980, Random House, New York. An overview of astronomy, with consideration of SETI.

Shklovsky, I.S., and Sagan, Carl, *Intelligent Life in the Universe*, 1968, Dell Publishing, New York. The classic Russian-American book that helped bring respectability to SETI. Semi-technical.

Sullivan, Walter, *We Are Not Alone*, 1966, McGraw-Hill, New York. The early days of SETI. Nontechnical.

Chapter 2. Martians and their Friends
Aliens in History

Dick, Steven J., *Plurality of Worlds*, 1967, Cambridge University Press, New York. Ancient ideas.

Hoyt, William Graves, *Lowell and Mars*, 1976, University of Arizona Press, Tucson. The life of Percival Lowell.

Chapter 3. Bug-Eyed Monsters and Little Green Men
Aliens in Science Fiction

Bradbury, Ray, *Martian Chronicles*, 1985 reprint, Bantam, New York. The classic collection of stories about the colonization of Mars.

Burroughs, Edgar Rice, *A Princess of Mars*, 1980 reprint, Del Rey, New York. The first adventures of John Carter and Dejah Thoris.

Doyle, Arthur Conan, *The Best Science Fiction of Arthur Conan Doyle*, 1981, Southern Illinois University Press, Carbondale, IL. Includes "The Horror of the Heights."

Franklin, H. Bruce, *Future Perfect*, 1978, rev. ed., Oxford University Press, London. American science fiction in the nineteenth century.

Gunn, James, *Alternate Worlds*, 1975, A & W Visual Library and Prentice Hall, Englewood Cliffs, NJ. A heavily illustrated history of science fiction.

Knight, Damon, *Science Fiction of the 30's*, 1975, Avon, New York. A collection selected by the distinguished science-fiction writer/critic.

Kyle, David, *Science Fiction*, 1976, Hamlyn Publishing Group, London. Heavily illustrated.

Nicholls, Peter (ed.), *The Science Fiction Encyclopedia*, 1979, Dolphin Books, New York. The definitive reference to the field.

Shelley, Mary, *Frankenstein*, 1965 reprint of the original novel, New American Library, New York.

Brooks, Jim, *Origins of Life*, 1985, Lion Publishing, Belleville, MI. Beautifully illustrated, nontechnical.

Bylinsky, Gene, *Life in Darwin's Universe*, 1981, Doubleday, Garden City, NY. Fascinating pictures of life as it might exist elsewhere.

Cairns-Smith, *Seven Clues to the Origin of Life*, 1985, Cambridge University Press, New York. The possibility that life arose on clay. Nontechnical.

Crick, Francis, *Life Itself*, 1981, Simon and Schuster, New York. The controversial theory that life did not arise here, by the codiscoverer of DNA.

Day, William, *Genesis on Planet Earth*, 1984, Yale University Press, New Haven. The details of the origin of life. Technical.

Goldsmith, Donald, *Nemesis*, 1985, Walker, New York. The theory that the dinosaurs were killed off by comets or meteorites. Nontechnical.

Gribben, John, *Genesis*, 1981, Dell Publishing, New York. The origin of life and the universe. Nontechnical.

Hoyle, Fred, and Wickramasinghe, N. C., *Diseases from Space*, 1979, Harper and Row, New York. The controversial theory that diseases arise in outer space.

Hoyle, Fred, and Wickramasinghe, N. C., *Evolution from Space*, 1981, Simon and Schuster, New York. The controversial theory that life did not arise here.

Hoyle, Fred, and Wickramasinghe, N. C., *Lifecloud*, 1978, Harper and Row, New York. The controversial theory that life arose in space.

Jonas, Doris and David, *Other Senses, Other Worlds*, 1978, Stein and Day, New York. Speculation about life on other worlds.

Margulis, Lynn, and Sagan, Dorion, *Microcosmos*, 1986, Summit Books, New York. The evolution of life. Nontechnical.

Ponnamperuma, Cyril, *The Origins of Life*, 1972, E. P. Dutton, New York. Heavily illustrated, nontechnical.

Sagan, Carl, *The Dragons of Eden*, 1977, Random House, New York. Speculations about the evolution of intelligence.

Scharf, David, *Magnifications*, 1977, Schocken Books, New York. Startling, exquisite scanning-electron-microscope pictures of the world around us.

Shapiro, Robert, *Origins*, 1986, Summit Books, New York. Nontechnical, critical examination of conflicting theories.

Chapter 5. Planet Trek
How to Find Planets

Black, David C. (ed.), *Project Orion*, SP-436, 1980, NASA, Washington, D.C. How to detect planets around other stars. Technical.

Dole, Stephen H., *Habitable Planets for Man*, 2nd ed., 1970, American Elsevier, New York. What kinds of planets are possible. Technical.

Chapter 6. Buck Rogers SETI
Exploring the Solar System

Beatty, J. Kelly, O'Leary, Brian, and Chaikin, Andrew (eds.), *The New Solar System*, 2nd ed., 1982, Cambridge University Press, New York. Semi-technical.

Cooper, Henry S. F., Jr., *The Search for Life on Mars*, 1980, Holt, Rinehart and Winston, New York. The story of the *Viking* landers.

Horowitz, Norman, *To Utopia and Back*, 1986, W. H. Freeman, San Francisco. The search for life in the solar system, by a *Viking* experimenter. Semi-technical.

Moore, Patrick, and Hunt, Garry, *Atlas of the Solar System*, 1983, Rand McNally, Chicago. Beautifully illustrated.

Murray, Bruce, Malin, Michael C., and Greeley, Ronald, *Earthlike Planets*, 1981, W. H. Freeman, San Francisco. A comparison of Mercury, Venus, Earth and Mars, with some discussion about exploring Mars in the future. Technical.

Washburn, Mark, *Mars at Last!* 1977, G. P. Putnam's Sons, New York. The story of the *Viking* landers.

Chapter 7. Sputnik and E.T.
Modern SETI

Ehrlich, Paul R., Sagan, Carl, Kennedy, Donald, and Roberts, Walter Orr, *The Cold and the Dark*, 1984, W. W. Norton & Co., New York. The Nuclear Winter, a possible determining factor in the Drake equation.

Chapter 8. Interstellar Postcards
Messages to Space

Sagan, Carl, Drake, F. D., Druyan, Ann, Ferris, Timothy, Lomberg, Jon, and Sagan, Linda Salzman, *Murmurs of Earth*, 1978, Ballantine Books, New York. The story of the *Voyager* message to the stars.

Chapter 9. Little Green Men and Nobel Prizes
False Alarms

Clark, David H., *Superstars*, 1984, McGraw-Hill, New York. Supernovas, pulsars and such. Nontechnical.

Greenstein, George, *Frozen Star*, 1983, New American Library, New York. Black holes and pulsars. Nontechnical.

Mitton, Simon, *The Crab Nebula*, 1978, Charles Scribner's Sons, New York. Technical.

Chapter 11. The Russians Are Looking, the Russians Are Looking!
International SETI

Tsiolkovsky, K.E., *Selected Works*, 1968, Mir Publishers, Moscow. Some of his pioneering technical papers on astronautics, with a biography.

Tsiolkovsky, K., *The Call of the Cosmos*, n.d., Foreign Languages Publishing House, Moscow. A modern translation of the rocket pioneer's science fiction.

Chapter 13. Beam Me Up, Mr. Spock!
Interstellar Travel

Finney, Ben R., and Jones, Eric M. (eds.), *Interstellar Migration and the Human Experience*, 1985, University of California Press, California. An extraordinary assortment of views from anthropologists to astronomers on space colonization and human history.

Forward, Robert L., "Roundtrip Interstellar Travel Using Laser-Pushed Lightsails," Vol. 21, No. 2, 1984, *Journal of Spacecraft*. Technical.

Forward, Robert L., "Starwisp: An Ultra-Light Interstellar Probe," Vol. 21, No. 2, 1984, *Journal of Spacecraft*. Technical.

Matloff, Gregory L., and Ubell, Charles, "World Ships: Prospects for Non-Nuclear Propulsion and Power Sources," Vol. 38, pp. 253-261, 1985, *Journal of the British Interplanetary Society*. A design for a thousand-year interstellar colony-ship. Technical.

Chapter 14. Are They Here?
UFO's and Other Evidence

Condon, Edward U. (Director), *Scientific Study of Unidentified Flying Objects*, 1969, Bantam Books, New York. Massive, critical study.

Gardner, Martin, *Fads and Fallacies in the Name of Science*, 1957, Dover Publications, Mineola, NY. Delightful exposé of hoaxes and myths, including UFO's.

Hynek, J. Allen, *The Hynek UFO Report*, 1977, Dell Publishing, New York. Hynek's view that UFO's should be taken seriously.

Klass, Philip J., *The Public Deceived*, 1983, Prometheus Books, Buffalo. Exposé of UFO fakery.

Klass, Philip J., *UFO's Explained*, 1976, Random House, New York. Reasonable explanations for many important UFO cases.

Kusche, Larry, *The Bermuda Triangle Mystery—Solved*, 1986, Prometheus Books, Buffalo. Why the Bermuda Triangle does not exist.

Oberg, James E., "Unidentified Fraudulent Objects," November, 1976, *Analog*. The UFO's that astronauts saw, identified.

Sheaffer, Robert, *The UFO Verdict*, 1981, Prometheus Books, Buffalo. A critical study of UFO's.

Story, Ronald, *The Space-Gods Revealed*, 1976, Harper & Row, New York. Why von Däniken's *Chariots of the Gods* and the like are nonsense.

Chapter 15. Where Have All the E.T.'s Gone?
Scientists Against SETI

Barrow, John D., and Tipler, Frank J., *The Anthropic Cosmological Principle*, 1986, Oxford University Press, New York. A detailed, scholarly argument that the universe was built for humans and that we may, after all, be alone.

Chapter 16. The Final Frontier
The Future

Bell, Trudy E., *Upward: Status Report and Directory of the American Space Interest Movement, 1984-85*, available from the author at 11 Riverside Dr., #15GW, New York, NY 10023. The history and directory of pro-space groups.

Bova, Ben, *The High Road*, 1983, Pocket Books, New York. Our future in space.

Contact, Jim Funaro, Dept. of Anthropology, Cabrillo College, Aptos, CA 95003. Annual conference of scientists, science-fiction people and the public, on First Contact. Try it!

Cooper, Henry S. F., *A House In Space*, 1976, Holt Rinehart Winston, New York. The story of our first space station, *Skylab*, and what it feels like to live in space.

Cultural Futures Research, Dept. of Anthropology, Box 15200, Northern Arizona Univ., Flagstaff, AZ 86011. Quarterly journal of future-oriented anthropological studies.

Haskins, Charles Homer, *The Renaissance of the 12th Century*, 1961, Meridian Books, World Publishing Co., New York. Culture shock, medieval style.

Heppenheimer, T. A., *Toward Distant Suns*, 1979, Fawcett Columbine, New York. One of the leading scientists in the field of space colonization tells how to do it.

Joels, Mark Kerry, *The Mars One Crew Manual*, 1985, Ballantine Books, New York. Semifactual handbook for an expedition to Mars.

Lovelock, James, and Allaby, Michael, *The Greening of Mars*, 1984, St. Martin's/Marek, New York. Colonizing Mars.

Lunan, Duncan, *Man and the Planets*, 1983, Ashgrove Press, Bath, England. Imaginative look at space exploration and colonization.

Maruyama, Magoroh, and Harkins, Arthur (eds.), *Cultures Beyond the Earth*, 1975, Random House, New York. Extraterrestrial anthropology.

Oberg, J. E., *Mission to Mars*, 1982, New American Library, New York. How we may go there.

Oberg, J. E., *New Earths*, 1981, Stackpole Books, Harrisburg, PA. How we can rebuild planets to make them nice places to live.

Oberg, J. E., and Oberg, Alcestis R., *Pioneering Space*, 1986, McGraw-Hill, New York. Our future on the final frontier.

O'Neill, Gerard, *The High Frontier*, 1977, William Morrow, New York. The book that brought scientific respectability to the idea of space colonization.

Stine, G. Harry, *The Third Industrial Revolution*, 1979, Ace Publishing, New York. How and why we will put industry in space.

Organizations of Interest

Astronomical League, Box 12821, Tucson, AZ 85732. For amateur astronomers.

Astronomical Society of the Pacific, 1290 24th Ave., San Francisco, CA 94122. For amateurs and pros, even those far from the Pacific.

British Interplanetary Society, 27/29 South Lambeth Rd., London SW8 1SZ, England. For amateurs and pros.

Campaign for Space, Box 1526, Bainbridge, GA 31717. Political Action Committee.

Citizens for Space, c/o Bill Simon, 3633 Empire Dr., Los Angeles, CA 90034. Political Action Committee.

Committee for the Scientific Investigation of Claims of the Paranormal, 3151 Bailey Ave., Buffalo, NY 14215-0229. Critically investigates UFO's and such.

L-5 Society, 1060 E. Elm, Tucson, AZ 85719. Supports human space flight and colonization.

National Space Society, West Wing Ste. 203, 600 Maryland Ave. SW, Washington, DC 20024. Among other things, sponsors Dial-a-Shuttle, a number to call when a Space Shuttle flight is orbiting.

Ohio State University Radio Observatory, Dept. of Astronomy, Ohio State University, Columbus, OH 43210. Needs money to continue the world's longest-running SETI system.

The Planetary Society, 65 N. Catalina Ave., Pasadena, CA 91106. Supports planetary exploration and SETI. The largest private space-group in the solar system.

Royal Astronomical Society of Canada, 124 Merton St., Toronto, Ontario M4S 2Z2, Canada. Very useful annual handbook.

SETI Institute, 101 First St. #410, Los Altos, CA 94022. Sponsors SETI research, including work on the NASA project.

Space Coalition, c/o Dickstein, Shapiro & Morin, 2101 L St., NW, Washington, DC 20037. Political Action Committee.

Spacepac, Suite S, 2801-B Ocean Park Blvd., Santa Monica, CA 90405. Political Action Committee.

Space Studies Institute, Box 82, Princeton, NJ 08540. Founded by space-colonization leader Gerard O'Neill, it supports colonization-oriented research projects.

U.S. Space Education Assn., 746, Turnpike Rd., Elizabethtown, PA 17022. International nonprofit, nonpartisan group promoting peaceful space exploration.

World Space Foundation, Box Y, So. Pasadena, CA 91030. Supports asteroid search and solar-sail research.

Write Now! Box 36851, Los Angeles, CA 90036-0851. Free information on how to encourage new space programs, such as by learning which politicians to write to. Practical and effective. (Enclose self-addressed, stamped envelope.)

INDEX